Collins

Key Stage 3 Science

Teacher Pack 3

Series editor: Ed Walsh
Authors: Tracey Baxter
Sunetra Berry
Pat Dower
Ken Gadd
Anne Pilling

William Collins' dream of knowledge for all began with the publication of his first book in 1819. A self-educated mill worker, he not only enriched millions of lives, but also founded a flourishing publishing house. Today, staying true to this spirit, Collins books are packed with inspiration, innovation and practical expertise. They place you at the centre of a world of possibility and give you exactly what you need to explore it.
Collins. Freedom to teach

Published by Collins
An imprint of HarperCollins*Publishers*
77 – 85 Fulham Palace Road
Hammersmith
London
W6 8JB

Browse the complete Collins catalogue at
www.collins.co.uk

10 9 8 7 6 5 4 3 2 1

ISBN 978-0-00-754024-2

Tracey Baxter, Sunetra Berry, Pat Dower, Ken Gadd and Anne Pilling assert their moral rights to be identified as the authors of this work.

British Library Cataloguing in Publication Data
A Catalogue record for this publication is available from the British Library

Commissioned by Letitia Luff
Project managed by Jane Roth
Series editor Ed Walsh
Managing editor Caroline Green
Project editor Amanda Redstone
Editorial assistance by Lucy Roth
Edited by Phillipa Allum, Elizabeth Barker, Hugh Hillyard-Parker, Sophia Ktori, Kate Redmond, Ros Woodward
Proofread by Tony Clappison, Sophia Ktori
Illustrations by Ken Vail Graphic Design
Typesetting by Ken Vail Graphic Design
Cover design by Angela English
Cover photograph by Riyazi/Shutterstock
Picture research by Joe Walter
Production by Rachel Weaver
Printed and bound by RR Donnelley at Glasgow, UK

Health and Safety: All practicals in this book have been safety checked (but not trialled) by CLEAPSS (www.cleapss.org.uk) and the publisher and authors have made every reasonable effort to ensure that the experiments and activities in this book are safe when conducted as instructed. However, the publishers assume no responsibility for any damage or injury caused or sustained while performing the experiments or activities in this book to the full extent permitted by law.

Parents, guardians, and/or teachers should supervise students who undertake the experiments and activities in this book, and carry out full risk assessments before carrying out any practical work.
CLEAPSS is an advisory service providing support in science and technology for a consortium of local authorities and their schools including establishments for pupils with special needs.

Acknowledgements
Every effort has been made to trace copyright holders and to obtain their permission for the use of copyright materials. The publishers will gladly receive any information enabling them to rectify any error or omission at the first opportunity.

Contents

Extra resources on CD-ROM

All pages from this print book available as PDF and editable Word files

Student worksheets in 2 formats; reusable and write-on (all available as PDF and Word files)

Practical sheets (as PDF and Word files)

Technician's notes (as PDF and Word files)

Each type of resource sheet collated in chapter batches to enable easy printing

Introduction

Purpose of the Collins Key Stage 3 Science course

This course has been developed to provide support to teachers in planning and delivering exciting, engaging and effective lessons. The overarching priorities have been to:

- ensure that the requirements of the National Curriculum Programme of Study have been responded to
- support teachers in teaching lessons in which students make good progress and are on track to achieve well at the end of KS4
- offer ways of tracking, reporting on and responding to progress given the move away from levels in the National Curriculum
- enable schools to select the period of time they decide to devote to KS3 before starting on KS4 courses
- address the need to challenge and engage students working at different levels of attainment
- focus on the development of skills and processes as well as content
- provide teachers with clear guidance as to how learning can be managed during the lesson, from initial engagement to consolidation and application.

Overview of changes at Key Stage 3

Some concepts previously associated with GCSE are now in KS3, such as:

- the role played by Watson, Crick, Wilkins and Franklin in the development of the DNA model
- the chemical properties of metal and non-metal oxides with respect to acidity
- the properties of ceramics, polymers and composites
- heating and thermal equilibrium
- the superposition of waves
- resistance as the ratio of potential difference to current
- the concept of electrostatic fields
- the light year as a unit of astronomical distance.

How Science Works has been replaced with Working Scientifically, which covers a similar range of skills and processes but which moves away from the social and economic implications of scientific developments.

There are no levels or any descriptors of progression. Nevertheless, schools are expected to track progress, report accordingly and respond with further challenge or support.

Organisation of the Collins Key Stage 3 Science course

There are two biology, two chemistry and two physics chapters in each year. This means that science can be delivered as one subject or by subject specialists. A scheme of work shows both 3-year and 2-year routes though the course to aid long-term planning, see page 198. Support and strategies for building and assessing progress are embedded throughout. For more details please see *Planning the learning process* on pages 3-7.

How the lesson plans work

Chapter introductions

These:

- give an overview of the of the content and skills covered in the chapter
- help in assessing prior learning and identifying misconceptions
- list the overarching learning objectives to help medium-term planning.

Learning objectives and outcomes

- Learning objectives for each topic are shared with students in the Student Book for short-term planning.
- Learning outcomes at three levels are listed, and it is shown which learning activities contribute to achieving each outcome.

These help in tracking the progress of all students, identifying those making good progress and who are on the right trajectory for sound results at GCSE, those who are making better than expected progress and are headed for the top results at GCSE, and those making less progress and for whom some intervention is needed.

Skills development

Ensuring progression in skills has underpinned the development of the course. There are three skill sets used, developed from a range of sources including PLTS, SEAL and APP in Science:

- **Thinking scientifically** relates to the relationship between evidence, ideas and theories, and has a strong role to play in science generally and in KS4 courses. It includes asking questions, considering the quality of evidence, understanding how theories develop, evaluating risks, using units and nomenclature, using equations and analysing data.
- **Working scientifically** relates to conducting practical investigations and includes making predictions, designing investigations, recording evidence, presenting evidence, interpreting evidence, developing explanations and evaluating data.
- **Developing as learners** is not a science-specific set of skills, though science can play a strong part in developing the skills, and they are recognised by the Ofsted school inspection framework. They include planning progress, acting responsibly, developing resilience, asking questions, communicating effectively, respecting others and collaborating effectively.

Resources needed gives an overview of all the resources needed for a lesson. More detailed guidance for practicals is given in student practical sheets and technician's notes on the accompanying CD-ROM.

Common misconceptions highlight specific misconceptions for the topic.

Key vocabulary is highlighted throughout to support literacy.

Teaching sequence

The lesson plans all use the same learning sequence. This is based on the idea that learning develops in stages during a lesson and that different parts of the lesson have different functions.

Engage This section draws students in to thinking about the ideas, and includes possible starter activities. Here students encounter ideas that will make them want to find out more.

Challenge and develop Students meet something that will challenge their existing understanding. It might be questions, ideas, demonstrations or experiments that make them realise the inadequacy of a simpler explanation.

Explain Students are encouraged to develop a good explanation and supported in capturing ideas in words or graphically. Differentiation ideas are given for students making less or more than expected progress.

Consolidate and apply Students realise how the new learning is to be consolidated and applied, including real-world applications. Again, differentiation ideas are given for students making different levels of progress.

Extend Addresses how the ideas of the topic can be extended to stretch students able to progress further.

Plenary suggestions Varied activities help in gauging student progress.

Answers All answers to Student Book questions and worksheet questions are provided.

Extra resources on the accompanying CD-ROM

Every lesson has an associated differentiated worksheet to support written work. These are provided as write-on versions with lines for student responses, and reusable versions without lines that can be used again and again to reduce photocopying costs. Practical sheets are provided to give support for planning, carrying out and analysing practical work. Technician's notes are provided to explain the materials and setup and help with planning.

Collins Connect is a linked publication of digital resources to enhance lessons. The resources are listed by lesson in the scheme of work on page 198 and in each lesson plan.

Planning the learning process: The thinking behind Key Stage 3 Science

Outstanding lesson plans

Each of the lesson plans is designed to support the delivery of high quality interactive lessons. They have a common structure that is transparent, making it easy for teachers to see what is suggested and make decisions about whether to use the plan as it stands or to customise. Sequences of lessons are planned to cover the requirements of the Programme of Study for Science (2013).

Objectives and outcomes

Each of the statements from the Programme of Study has been built into the plan and identified as 'overarching objectives'. From these authors have developed learning objectives for individual lessons; these draw upon processes as well as content. These objectives use the stem of 'we are learning how to' as it is important that students develop abilities as opposed to just learning facts.

From these, outcomes were developed; there is a strong relationship between objectives and outcomes. Outcomes use the stem 'what I'm looking for' (in which 'I' is the teacher) and indicate the evidence the teacher should see that indicates that learning has occurred.

The outcomes also play a critical role in the **assessment of progress**. The 2013 Programme of Study has no level descriptors but the importance of tracking progress hasn't diminished. Part of the function of the outcomes is to enable the teacher to decide if students are making good progress. The outcomes are differentiated; students who are making good progress will be, generally speaking, achieving at least two out of the three outcomes. Students regularly achieving all three may need, on occasions, more challenging activities. Those struggling to achieve more than one may need additional support. However the Teacher Pack materials are designed to be used in a flexible way, to be selected from and adapted to meet the needs of students.

Skills

Each lesson has two or three focus skills selected. The three sets of skills are as follows:

- **Thinking scientifically**. These are the skills that scientists use in a variety of contexts, such as asking questions, using equations and analysing data. Although these are used in practical activities their role is much wider than that. All lessons have such a skill identified; this skill will also be reflected in the learning activities and often feature in the objectives and outcomes.
- **Working scientifically**. These skills are strongly based on the 'Working Scientifically' aspect of the Programme of Study and are selected where practical activities form part of the lesson. Although students may well be developing and using a number of these skills there is always one in particular that is highlighted. This is designed to make it easier for teachers and students to focus upon particular aspects. Over a sequence of lessons the focus will change so that all are covered over a period of time. Obviously the teacher can select a different focus if their view is that this better serves students' learning needs. However what is less effective is a lesson in which a range of skills are ostensibly targeted but what happens in practice is that the same few (often gathering data and writing conclusions) are emphasised repeatedly.
- **Learner development**. Science has the power to develop a wider range of skills than those which are subject specific; it also draws upon these wider skills. In every lesson one of these is focused upon and reflected in the learning activities, providing an opportunity for the lesson to be placed in the wider context of the development of the student as a learner.

None of these skills are radical or original in nature, either in their form or function. Schools familiar with APP, PLTS or SEAL will recognise many of the components. However what the writing team have done is to present these in a form that is both manageable and integrated with the delivery of content.

The learning cycle

The main part of each lesson plan forms a common sequence as part of a 'learning cycle'. The basic idea behind this can be traced back to an inquiry lesson planning model developed for the Science Curriculum Improvement Study (SCIS) program in the USA, a K-6 science program in the early 1970s. In its original form the learning cycle model had 3 stages (exploration, invention and discovery). The idea was that the teacher would develop the science concept in the second stage, rather than defining it at the outset of the lesson as in the traditional approach. The introduced concept subsequently enables students to incorporate their exploration in the third stage and apply it to new examples. It thus has a strong link to a constructivist model of learning, in which students are challenged to develop ideas rather than being presented with a body of content to accept 'as is'. Many examples of learning cycles have since been developed; one current version is the 5E Learning Cycle, which is used in the new BSCS (Biological Sciences Curriculum Study) science programmes as well as in other texts and materials. The Collins KS3 Science sequence consists of the following stages:

- **Engage**, in which the teacher draws students into thinking about certain ideas. Here a student encounters ideas that will make them want to find out more.
- **Challenge and develop**, in which a student meets something that will challenge their existing understanding. It might be questions, ideas, demonstrations, experiments, etc., that make them realise the inadequacy of a simpler explanation.
- **Explain**, in which students are encouraged and supported to develop a good explanation. This draws upon the previous section and supports students to capture ideas in words or graphically.
- **Consolidate and apply**, where students realise how the new learning is to be consolidated and applied. Real world applications are often drawn upon at this point.
- **Extend**, in which ideas can be extended by students making greater than expected progress.

Throughout, the emphasis is upon students' active learning. Authors have identified what students should be doing to engage with each phase, and the purpose of the component, as well as the nature. This is designed to present the teacher with an open, easily assimilated structure which shows what is intended at each point. A teacher wishing to substitute their own activity will therefore be supported in knowing what it should do to support progress. Similarly a teacher using the plan as it is can see the reasoning behind what they are doing.

Encouraging discussion

Discussion is important for exploring ideas and evidence in science and for developing an understanding of concepts, as well as improving literacy. There is a wide variety of ways of organising small-group discussion. Eight of these are used extensively in the lesson plans:

- **Pair talk** This is easy to organise even in cramped classrooms. It is also ideal to promote high levels of participation and to ensure that the discussions are highly focused, especially if allied to tight deadlines. It can be used in the early stages of learning for students to recall work from a previous lesson, generate questions, work together to plan a piece of writing or take turns to tell a story. Pairs can also be used to work as reading partners with an unfamiliar text. It is also ideal for quick-fire reflection and review, and for rehearsal of ideas before presenting them to the whole class.
- **Pairs to fours** Students work together in pairs – possibly friendship, possibly boy/girl, etc. Each pair then joins up with another pair to explain and compare ideas.
- **Listening triads** Students work in groups of three. Each student takes on the role of talker, questioner or recorder. The talker explains something, or comments on an issue, or expresses opinions. The questioner prompts and seeks clarification. The recorder makes notes and gives a report at the end of the conversation. Another time roles are changed.
- **Envoys** Once groups have carried out a task, one person from each group is selected as an 'envoy' and moves to a new group to explain and summarise, and to find out what the new group thought, decided or achieved. The envoy then returns to the original group and feeds back. This is an effective way of avoiding tedious and repetitive 'reporting back' sessions. It also puts a 'press' on the envoy's use of language and creates groups of active listeners.
- **Snowball** Individuals explore an issue briefly; then pairs discuss the issue or suggest ideas quickly; then double up to fours and continue the process into groups of eight. This allows for comparison of ideas, or to sort out the best, or to agree on a course of action. Finally, the whole class is drawn together and a spokesperson for each group of eight feeds back ideas.
- **Rainbow groups** This is a way of ensuring that students are regrouped and learn to work with a range of others. After small groups have discussed together, students are given a number or colour. Students with the same number or colour join up, making groups comprising representatives of each original group. In their new group, students take turns to report back on their group's work and perhaps begin to work on a new,

combined task.

- **Jigsaw** A topic is divided into sections. In 'home' groups of four or five, students allocate a section each, and then regroup into 'expert' groups. In these groups, they work together on their chosen area, then return to original 'home' groups to report back on their area of expertise. The 'home' group is then set a task that requires the students to use the different areas of 'expertise' for a joint outcome. This requires advance planning, but is a very effective speaking and listening strategy as it ensures the participation of all students.
- **Spokesperson** Each group appoints a spokesperson. The risks of repetition can be avoided if:
  - one group gives a full feedback, and others offer additional points if not covered;
  - each group is asked in turn to offer one new point until every group 'passes';
  - groups are asked to summarise their findings on A3 sheets which are then displayed – the class is invited to compare and comment on them.

From: Literacy in Science, DfES © Crown copyright 2004

Starter and plenary toolkit

Authors have drawn on an varied selection of learning strategies, bringing ideas from lessons they have given or observed themselves, and drawing extensively from the following list:

- **Silent animation** Show students a short video clip without sound and ask them to say what they think is happening. They could either suggest the ideas or do a 'voice over' for it.
- **Ideas hothouse** Ask students to work in pairs to list points about what they know about a particular idea. Then ask the pairs to join together into 4s and then 6–8s to discuss this further and to come up with an agreed list of points. Ask one person from each group to report back to the class.
- **Where's the answer?** As students enter the room, give them a card with a word written on it. Then ask a question, ask them if they think they are the answer and why.
- **What's in the picture?** Have a complex picture such as the water cycle or a photo of an industrial process concealed by rectangles, each of which can be removed in turn. Ask for suggestions as to which tile should be removed and what the partially revealed graphic shows. Encourage speculation and inference; ask for suggestions as to which tile should go next and continue until a full explanation has been produced.
- **Learning triangle** At the end of a lesson ask students to draw a large triangle with a smaller inverted triangle that just fits inside it (so they have four triangles). In the outer three ask them to think back over the lesson and identify (and write in the respective triangles):
  - Something they've seen
  - Something they've done
  - Something they've discussed
  - Then to add in the central triangle: something they've learned
- **Heads and tails** Ask each student to write a question about something from the topic on a coloured paper strip and the answer on another colour. In groups of 6–8, hand out the strips so that each pupil gets a question and an answer. One student reads out their question. The student with the right answer then reads it out, followed by their question.
- **What do I know?** Ask students to each write down one thing about the topic they are sure of, one thing they are unsure of and one thing they need to know more about, being specific. Ask them to work in groups of 5–6 to agree on group lists, ask each group to say what they decided and agree as a class about what they are confident about, what they are less sure of and what things they want to know more about. This is a good way towards the end of a topic of identifying topics for revision.
- **Freeze frame** Ask students to create a 'freeze frame' of an idea in the topic. This involves 3–4 pupils arranging themselves as static image and other pupils to suggest what it represents.
- **The big ideas** Ask students to write down three ideas they learnt during the topic. Then ask them to share their facts in groups and to compile a master list of facts, with the most important at the top. Ask for ideas to be shared and find out which other group(s) agreed.
- **Ask me a question** Ask students to write a question about something from the topic and then a mark scheme for the answer. Encourage them to come up with ones worth more than 1 or 2 marks and to try out their questions on one another.
- **Hot seat** Ask each student to think up a question, using material from the topic. Select someone to put in the hot seat, ask students to ask their question and say at the end whether the answer is correct or incorrect.

Assessment of learning

There is a need for teachers to be able to assess the effectiveness of the learning and, indeed, for students to assess their own progress. This needs to be done for three reasons:

- Teachers need to gauge the effectiveness of a lesson, or sequence of lessons, to judge what the next steps should be.
- Student progress needs to be reported upon, both for accountability purposes but also to inform a whole school perspective on providing challenge and intervention.
- Teachers need to be able to assess whether the design of the lessons was effective for the next time they teach it.

This has, of course, been done previously using a system of levels. When the 2013 Programme of Study was published, it contained no level descriptors and the DfE indicated that it was dropping this calibration of progress. This doesn't mean that schools are not expected to assess progress, use it to plan next steps in learning or report on how individuals are doing. At the time of publication, levels are still being widely used, though this is likely to change as KS2 outcomes move to a new reporting system, GCSE outcomes move to a new set of grades and schools gain confidence in assessing progress.

Trajectories of Progress

Progress in KS3 Science can be represented by Trajectories of Progress based on what individual students need to achieve to be 'making good progress'. The three differentiated outcomes in each lesson plan (and the three coloured levels in the student book) correspond to three model trajectories for lower, average and higher attaining students.

An average student who studies the materials, encounters the ideas, tries the assessment activities and gains 40-70% of the marks is deemed to be progressing well. The materials enable the teacher to gather evidence to this effect, report accordingly and plan subsequent lessons effectively. Students gaining full (or close to full) marks may need additional challenge; those scoring half marks or under may need intervention.

However, schools may be using tracking systems still based on levels and it may be that data from the materials in these resources has to be used to support this. As a rule of thumb, assessment data from the books can be used to inform a levels based system as follows:

	Old NC level	KS3			KS4	
		Book 1	Book 2	Book 3	Yr 10	Yr 11
		10	10	10	10	A*
Student responses to broad range of assessment evidence, including Applying key ideas activities and end of topic Questions:		9	9	9	9	A
		8	8	8	8	B
		7	7	7	7	C
		6	6	6	6	D
Higher ToP: scores above 70%	**5**	5	5	5	5	E
Average ToP: scores 40–70%	**4**	4	4	4	4	F
Lower ToP: scores below 40%	**3**	3	3	3	3	G

FIGURE 1 Trajectory of Progress (ToP) the Collins Key Stage 3 Science progress tracking system

The Collins KS3 Science scheme has a number of assessment devices available to students and teachers. As well as questions in each spread which will give rise to written and spoken outcomes, there are also:

- **Ideas you have met before** summarises ideas that students will have met previously at the start of each chapter. These provide an opportunity for discussions around prior learning so that the teacher can gauge what is already understood and plan teaching accordingly.
- **Applying key ideas** sections near the centre of each chapter. These provide an opportunity for students and teachers to assess how well key ideas from the earlier part of the topic have been mastered. The Teacher Guide suggests how additional support can be provided if responses are not as good as expected.

- **Checking progress** sections at the end of each chapter showing how the key ideas progress. This supports students in assessing their own progress and targeting areas they need to focus on. The teacher could get students to suggest why they've assessed themselves as they have.
- **Questions** at the end of each chapter, with a balance of style and focus. The styles of questions include objective test questions, short written answers and longer responses. The focuses include knowledge & understanding, application and evaluation. This means that responses can be analysed against these criteria as well as against area of content.

1.1 Introduction

When and how to use these pages

The Introduction in the Student Book indicates some of the ideas and skills in this topic area that students will already have met from KS2 or from previous KS3 work, and provides an indication of what they will be studying in this chapter. *Ideas you have met before* is not intended to comprehensively summarise of all the prior ideas, but rather to point out a few of the key ones and to support the view that scientific understanding is progressive. Even though students might be meeting contexts that are new to them, they can often use existing ideas to start to make sense of them.

In this chapter you will find out indicates some of the new ideas that the chapter will introduce. Again, it isn't a detailed summary of content or even an index page. Its purpose is more to act as a 'trailer' and generate some interest.

The outcomes, then, will be recognition of prior learning that can be built on, and interest in finding out more.

There are a number of ways this can be used. You might, for example:

- Use *Ideas you have met before* as the basis for a revision lesson as you start the first new topic.
- Use *Ideas you have met before* as the centre of spider diagrams, to which students can add examples, experiments they might have done previously or what they found interesting.
- Make a note of any unfamiliar/difficult terms and return to these in the relevant lessons.
- Use *In this chapter you will find out* to ask students questions such as:
  - Why is this important?
  - How could it be used?
  - What might we be doing in this topic?

Overview of the chapter

In this chapter, students will learn about the variation between and within species, the causes and types of variation and the importance of biodiversity. They will learn how natural selection operates and that we artificially select features we favour in organisms, but that this can upset the balance of nature and leave organisms vulnerable to natural selection processes if their environment changes. The students will discover how scientists developed the current international system for naming and classifying organisms, as well as how scientists compete and collaborate to make new discoveries – for example the discovery of the structure of DNA. They will explore the structure of chromosomes, genes and DNA, and will investigate their importance in the inheritance of characteristics.

This chapter offers a number of opportunities for the students to investigate the use of models to help develop explanations and to research and evaluate evidence to construct explanations. These are explored in activities throughout the chapter.

Obstacles to learning

The students may need extra guidance with the following terms and concepts:

- **Inheritance** Many students think that acquired characteristics may be passed on to offspring. Chromosomes need to be visualised for students to understand how small they are – this will help them to understand the relative size and complexity of genes and DNA. Some students may think that freckles are a disease. Another common misconception is that parents' traits are mixed in the offspring – for example if the mother has curly hair and the father has straight hair, then the children will all have wavy hair.
- **Natural selection** The concept of natural selection is difficult. Many students think that natural selection is a random process and that evolutionary change in a species can happen during a lifetime.
- **Identical twins** The students need to understand that identical twins are not always exactly the same. Stress the importance of environmental factors in creating variation.

- **Chromosomes and genes** The inheritance of gender is not fully covered – if you do choose to teach it more fully stress the fact that it involves whole chromosomes and not just a gene. The concept of dominant/recessive genes is very hard for students to grasp.

	Topic title:	Overarching objectives:
2	Exploring differences	Differences between species The importance of maintaining biodiversity
3	Looking more closely at variation	The variation between individuals within a species being continuous or discontinuous, to include measurement and graphical representation of variation
4	Exploring the causes of variation	Heredity as the process by which genetic information is transmitted from one generation to the next
5	Learning about selective breeding	
6	Finding out how organisms survive	The variation between species and between individuals of the same species means some organisms compete more successfully, which can drive natural selection
8	Understanding why siblings are different	Heredity as the process by which genetic information is transmitted from one generation to the next
9	Looking inside a cell's nucleus	A simple model of chromosomes, genes and DNA in heredity, including the part played by Watson, Crick, Wilkins and Franklin in the development of the DNA model
10	Learning about DNA	
11	Exploring human chromosomes	
12	Understanding cloning	
13	Explaining extinction	Changes in the environment may leave individuals within a species, and some entire species, less well adapted to compete successfully and reproduce, which in turn may lead to extinction The importance of maintaining biodiversity and the use of gene banks to preserve hereditary material

1.2 Exploring differences

Lesson overview

Learning objectives

- Identify the differences between different species.
- Explain the importance of diversity.

Learning outcomes

- Recall differences between different species and describe how we use variation to classify organisms. [O1]
- Explain the importance of biodiversity and present the key points. [O2]
- Explain how the similarities and differences between organisms can be used to classify them. [O3]

Skills development

- Thinking scientifically: analyse data
- Working scientifically: interpret evidence
- Learner development: collaborate effectively

Resources needed a selection of leaves with different shapes, vein patterns, margins etc.; Worksheet 3.1.2a; Worksheet 3.1.2b (copied onto card)

Digital resources Quick starter; Slideshow: The five kingdoms - protoctists, prokaryotes, fungi, plants and animals; Interactive activity: Drag the animals to the correct phyla; Slideshow: Life in different environments - A look at how living things adapt to extreme environments; Slideshow: Hybrid - Definition and examples; Video

Key vocabulary species, variation, biodiversity, ecosystems, classification

Teaching and learning

Engage

- **Think, pair, share** Ask the students, working in pairs, to **discuss** the question 'How many different types of living things are there on Earth?'. Take feedback and then ask student pairs to read the 'Variety of life' section in the Student Book and **discuss** questions 1, 2 and 3. [O1]
- Still working in pairs, the students complete task 1 of Worksheet 3.1.2a and then select pairs to **share** their work. Discuss and develop these responses with the class. [O1]

Challenge and develop

- Give each pair the 'leaf sort' cards from Worksheet 3.1.2b. Ask them to **classify** the leaves into groups. If possible, display some real leaves that cover all the variables on the worksheet. Take feedback about how the students grouped the leaves. Emphasise the similarities and the differences. [O1]
- In groups, the students **discuss** why scientists group animals together, how they decide what groups to place animals in and how they are named. Take feedback and discuss the importance of a universal naming system. [O1&3]

Explain

- **Group work** The students read and **discuss** the 'Why is biodiversity important?' section in the Student Book. Allocate each group a set of four animals to **research** (task 2 of Worksheet 3.1.2a). Each group **summarises** the common features and differences between the animals in their set. [O1&2]
- **Pair work** The students work in pairs to **explain** in no more than 20 words why variation is important. [O2]
- Student pairs **design** posters using pictures and diagrams, and as few words as possible, to show why biodiversity is important. If time allows, each pair can **present** and **explain** their poster to the class. They then complete task 2(b) on Worksheet 3.1.2a. [O2]

Give higher-attaining students a picture of a woodland habitat and ask them to **predict** the roles of different species within the ecosystem to exemplify the importance of biodiversity.

Consolidate and apply

- **Pairs to fours** Working in pairs, the students read the 'Classification' section in the Student Book and then, in fours, to **agree on** answers to questions 6 and 7. Take feedback, explaining that similarities between organisms are just as important as differences when classifiying a species. [O3]

Extend

- Ask students able to progress further to **discuss** and **predict** why scientists are more interested in the details of internal and external structures than the behaviour or general appearance of a vertebrate. [O3]

Plenary suggestions

Pair summaries The students work in pairs to **summarise** the lesson in no more than 30 words. Take feedback and let the whole group **peer-assess** the statements. [O1, 2&3]

The big ideas The students write down three ideas they learned during the lesson. They **share** their facts in groups and **compile** a prioritised list of facts. Take feedback and find out which other group(s) agreed with the prioritised points. [O1, 2&3]

Answers to Student Book questions

1. it is a group of organisms with similar features that can produce fertile young
2. *similarities:* they all have muscular bodies; are efficient hunters; have good vision, claws and sharp teeth

   differences: tigers are largest; have stripes; lions have a mane; are spotted at birth but spots disappear; have shorter fur; claws retract; the cheetah has larger nostrils; cannot retract claws; has solid spots; is smallest and fastest
3. the two small cats (Himalayan and Cymric) are of the same species; they have similar features and can breed together
4. the variety of organisms/plants and animals
5. ecosystems are more able to cope with changes, because there is a better chance of organisms having features that will enable them to survive
6. A universal classification system means that all scientists call organisms by the same name.
7. the new species would be classified according to its features; the scientists would then be able to predict relationships to other organisms and describe general internal features

Answers to Worksheet 3.1.2a

1. a) sharp canine teeth for ripping and tearing; claws; forward facing eyes

   b) i) *similarities*: fur; facial features; carnivores; claws; walk on toes ii) *differences*: shape of pupil (round in big cats; slits in most small cats); big cats roar, whereas small cats purr; any other correct observation

   c) variation
2. a) set A – *Common features*: any common mammal features; *differences*: size; colour; weight; life span; location; lemurs depend more on sense of smell but orangutan/chimps more on sight; lemurs have a tail; orangutans/chimps have hair instead of fur; lemurs have claw on second toe; marmosets have claws; chimps and orang utans have finger nails, opposable thumbs, higher brain-to-body size ratio, high level of intelligence, ability to grasp with fingers and/or toes, padded digits with fingerprints

   set B – *common features*: any common reptile features; *differences:* colour; size; location; snake has no legs; tortoise has shell and not scales; agamas can tolerate very high temperatures; iguana has spines running down its back.

   set C – *common features*: any common features of amphibians; *differences*: size; colour; location; African bullfrog females are smaller than males; salamanders have longer bodies

   set D – *common features*: any common features of birds; *differences*: colour; size; face shape; beak shape; diet; location; owls are nocturnal; macaw brightly coloured feathers; peacock has large tail; robin has red breast

   set E – *common features*: any common features of fish; *differences*: colour; size; location; shape; sharks have a large dorsal fin and skeleton made of cartilage; clown fish have symbiotic relationship with anemones; plaice are flatfish

   b) so that all scientists are clear what the organism is
3. a), b) A – *liger*: offspring of a male lion and a tigress; has faint stripes (and spots); this male has small mane and ruff

   B – *beefalo*: offspring of cow and buffalo; hind parts are like that of a cow; this male has mane and horns of a buffalo

   C – *zorse*: offspring of a female zebra and a male horse; body shape of a horse but with zebra stripes on legs and neck and more faintly on other parts of body

   D – *zedonk*: offspring of a zebra and a donkey; head and body shape of a donkey, but zebra stripes on legs and hair on neck

   c) ethical issues to be discussed include: hybrids are usually sterile; have short life spans; many genetic deformities

1.3 Looking more closely at variation

Lesson overview

Learning objectives

- Explain the difference between continuous and discontinuous variation.
- Investigate variation within a species.
- Evaluate the importance of variation in organisms.

Learning outcomes

- Explain the difference between continuous variation and discontinuous variation by looking at the differences in features between individuals of the same species. [O1]
- Identify variation within a species by gathering and presenting data about individuals within a species; explain the difference between variation within species and across species. [O2]
- Evaluate the importance of variation in organisms by assessing the advantages and disadvantages of variation to a species' survival. [O3]

Skills development

- Thinking scientifically: ask questions
- Working scientifically: present evidence
- Learner development: respect others

Resources needed at least 40 holly (or other) leaves in bags of 5, the bags labelled A, B, C etc.; tape measures; graph paper; spreadsheet software; Worksheet 3.1.3

Digital resources Quick starter; Interactive activity: Drag the statement to the correct correlation

Key vocabulary continuous variation, discontinuous variation, correlation, inbreeding

Teaching and learning

Engage

- Show the students photographs of the British Royal family (see task 1 of Worksheet 3.1.3). Ask them to **brainstorm** in groups: 'How are these people alike? How are they different?'. Take feedback. [O1]
- Ask the students to look at the pictures of *Cepaea nemoralis* snails in task 1 of Worksheet 3.1.3. Explain that these snails are all the same species. Ask the student pairs to **identify** how the snails vary. [O1]

Challenge and develop

- Ask the students to work in pairs to read the 'Types of variation' section in the Student Book and answer questions 1 and 2. [O1]
- **Small groups** Working in groups of three, the students **measure** one feature of members of the class. They **record** the data in a class spreadsheet, then complete parts a) and b) in task 2 of Worksheet 3.1.3. [O1&2]
- Ask the students to **observe** and **investigate** different individuals of the same species – for example whether or not longer holly leaves have more prickles. The students **predict** the outcome, **decide** how to collect data and a suitable sample size. Give each table one bag of leaves to investigate (see *Resources needed*). Students replace the leaves in the bag and pass to the next group. Repeat until each group has examined every bag of leaves. They **record** their data in a spreadsheet. [O1]
- The students work in pairs to **write an explanation** of the difference between variation within and across species. [O2]

Explain

- **Pair work** Still in their pairs, the students read the 'Investigating variation' section in the Student Book and answer questions 3 and 4. [O2]
- The pairs then use the spreadsheets of data from both investigations (class features and leaf features) to **produce graphs**. Ideally, these should be computer-generated. [O2]

  If graph-drawing software is not available, graphs with axes pre-drawn can be provided for lower-attaining students.
- The students then **peer-review** the graphs. Encourage them to **comment on** features such as labelling, use of an even scale, correct plotting, accuracy of line and size of graph. [O1&2]

Consolidate and apply

- The students **work independently** to complete part c) in task 2 of Worksheet 3.1.3 concerning variation in peaches. [O1&2]
- **Pair talk/Pairs to fours** Ask the students to read the 'Importance of variation' section in the Student Book. Working first in pairs, then fours, they **discuss** the question 'Is variation important, and why?'. [O3]
- Working in groups, the students **discuss** the question in task 3 of Worksheet 3.1.3. [O3]

Extend

Ask students able to progress further:

- to work in pairs to **suggest** and **investigate** correlations using the graphs produced in this lesson – for example, do taller students have wider arm spans? [O1&2]
- to work in groups to find some leaves growing in shade and also non-shade (e.g. from a large bush), **calculate** the areas of the leaves (e.g. blackberry leaves) taken from open and shaded habitats, and to **suggest** why variations occur. [O1, 2&3]

Plenary suggestions

Mind map Working in pairs, the students draw a mind map of what they have learned during the lesson. [O1,2&3]

What do I know? The students identify three things they have learned in this lesson, one thing they want to know more about and one target for future learning. [O1, 2&3]

Answers to Student Book questions

1. *dogs*: e.g. tail length (continuous); ear position (discontinuous); height (continuous); leg length (continuous); colour (discontinuous); *humans*: e.g. eye colour (discontinuous); ear shape (continuous); hair colour (discontinuous); height (continuous); foot size (continuous)
2. *Continuous*: the feature changes gradually over a range of values. *Discontinuous*: individuals have one form of a feature or another.
3. Steps include: randomly pick a sample of 20+ leaves, from same tree; measure their lengths; count the spikes; plot a scatter graph.
4. The larger the sample size, the more accurate the investigation.
5. Variation helps to ensure that when conditions change, some individuals will survive; it also helps to pass on the 'best features' to the next generation.
6. The brown squirrel has the advantage, because it is more camouflaged and will not be seen by predators, unlike the albino squirrel.

Answers to Worksheet 3.1.3

1. a) *similar features*: all female; all have fair skin; fairly wide mouth; fairly prominent nose b) gender: count; skin colour: count; mouth size: measure; nose size: measure; for any other features identified depends on the type of feature

   c) three of: colour of shell; size of shell; bands/no bands; thickness of bands; number of bands

   d) *colour* = count; *size* = measure; *bands/no bands* = count; *thickness of bands* = measure; *number of bands* = count
2. a) height; weight; eye colour; hair colour; skin colour; any suitable feature b) height and weight are continuous; eye colour, hair colour and skin colour are discontinuous c) i) line graph ii) 120–140 g iii) 143–157 g iv) variety 1 = 130 g; variety 2 = 149.5 g v) continuous
3. Variation gives a survival advantage to some organisms; it is important in reproduction when selecting a mate that will give offspring the best chance of survival; weaker individuals will not survive if the environment changes.

1.4 Exploring the causes of variation

Lesson overview

Learning objectives

- Identify some features of organisms that are inherited and some that are determined by the environment.
- Understand that offspring from the same parents may show considerable variation.
- Evaluate the importance of genetic and environmental variation to the survival of the organism.

Learning outcomes

- Recognise that features such as height, eye colour, freckles etc. are inherited, whereas scars, tooth loss, tattoos etc. are gained from the environment. [O1]
- Explain that offspring from the same parents may be very different because they have unique, random combinations of their parents' hereditary information. [O2]
- Evaluate the relative importance of genetic and environmental variation, and conclude that genetic variation is essential for long-term survival and environmental variation can affect short-term survival. [O3]

Skills development

- Thinking scientifically: design investigations
- Working scientifically: analyse data
- Learner development: collaborate effectively

Resources needed plant trays; oat seeds; soil and other growing mediums (e.g. vermiculite, compost, perlite); Worksheet 3.1.4; Practical sheet 3.1.4; Technician's notes 3.1.4

Digital resources Quick starter; Slideshow: The causes of variation - A look at genetic and environmental factors; Interactive activity: Drag the characteristics to the correct group - caused by genetic factors, environmental factors, or both; Video

Common misconceptions Acquired characteristics may be passed on to offspring.

Key vocabulary inherited, genetic

Teaching and learning

Engage

- Ask the students: 'What if we were all the same? What would happen if a new disease emerged that could kill us?'. Use this discussion to reinforce why variation is so important for survival. [O3]

Challenge and develop

- **Pair work** The students work in pairs to read the 'What causes variation?' section in the Student Book and answer questions 1 and 2. Take feedback and explore with the students how some features are both inherited and caused by the their environment – for example height. The students **discuss** and **write a list** of features that can be affected by the environment. [O1]
- The students work in their pairs to complete task 1 of Worksheet 3.1.4. [O1]
- **Pair talk/pairs to fours** Ask the question 'If you cut off a mouse's tail, will its young have no tails?'. Take feedback and stress that inherited variation passes from parents to offspring, but environmental variation is not passed on. [O1&2]

Explain

- As a class, discuss ways in which the students think they are similar to their parents, grandparents, brothers and sisters. Then ask them to work in groups to **brainstorm** why children from the same parents look different. Take feedback and discuss the students' ideas. [O1&2]

- Working in pairs, the students read the 'Why are offspring different?' section in the Student Book and answer questions 3 and 4. They can then complete task 2 of Worksheet 3.1.4. [O1&2]
- **Group work** The students **begin an investigation** of what affects the height of seedlings (see Practical sheet 3.1.4). Encourage them to **make decisions** about how often they will take measurements, how many seeds they will need in order to collect enough evidence and what their control will be. Ensure that they wash their hands after handling soil/compost. [O1]
- The students read the 'Genetic or environmental?' section in the Student Book. Ask them to work in pairs to answer question 5. Take feedback and discuss their ideas. [O2]

Consolidate and apply

- When the students have sufficient data of seedling growth (after two to three weeks), ask them to **consider** what the evidence shows. Discuss why seedling height varies (inherited and environmental factors). Ask whether the variation of seedling height in one condition is wider or less than the variation between different conditions. [O1, 2&3]
- **Small groups** The students work in threes to **design** a poster about genetic and environmental variation, **present** their work and **peer-review** the work of others. [O1, 2&3]
- **Pairs to fours** The students answer part a) of task 3 of Worksheet 3.1.4 independently, then **share** with a partner and join with another pair to **agree on** an answer. Take feedback and discuss their ideas. [O3]

Extend

Ask students able to progress further:

- to work in groups to **brainstorm** other examples of how the environment can affect variation in plants and animals – for example stoats change their coat colour in response to colder temperatures. [O1&3]
- to complete task 3 of Worksheet 3.1.4 to **evaluate** the importance of genetic and environmental variation in the survival of orangutans. [O3]

Plenary suggestions

Ask me a question The students work in groups to **devise questions** starting 'How', 'Where', 'When', 'Why' or 'What' on the topic, with answers. They **answer** another group's questions and **peer-assess**. [O1, 2&3]

Answers to Student Book questions

1. *inherited*: eye colour; shape and size of ears; etc.; *environmental*: skin colour; colour of nails (varnish); scars; tattoos; etc.
2. any of: coat colour; coat pattern; length of mane; length of tail
3. each child gets a set of genetic information from their father and a set from the mother; the combinations are different
4. a) straight hair; face shape; skin colour; eyebrow shape b) eye colour; shape of nose; mouth shape; thickness of eyebrows; hair colour
5. height; weight; hair colour
6. In the long term genetic variation is more important than environmental, because only genetic variation is passed on to offspring. In the short term the environmental variation may be more important if an organism is to survive, e.g. the coat colour of the stoat.

Answers to Worksheet 3.1.4

1. a) i) variation passed on from parents ii) variation caused by the environment that the organism lives in

   b) inherited: female; blue eyes; shoe size; can roll tongue; environmental: can speak French; good at tennis; has pierced ears; has a tan; both: height; blond hairv c) i) no ii) no; because environmental variations are not inherited
2. a) i) from the parents by genetic variation ii) amount of water; amount of light; amounts of different minerals

   b) They are different ages; they have inherited different combinations of information from their parents; they also have environmental differences, e.g. hair style.
3. a) there is no *inherited variation*, but the twins are affected differently by *environmental variation* – in this case ear piercing and hair cutting b) i) this results in less variation, which could mean that pandas will not survive long-term changes to their environment ii) this increases variation, so pandas are more likely to survive changes in their environment over the long term iii) Less food means that fewer pandas survive in the short term to breed and ensure the species' long-term survival. iv) More food means more that pandas are likely to survive in the short term, increasing long-term survival prospects. v) In the short term, many pandas may die; but the greater the variation between pandas, the more likely it is that enough will survive by having greater immunity to the new disease.

1.5 Learning about selective breeding

Lesson overview

Learning objectives

- Describe how selective breeding can produce organisms with desirable features.
- Explain the process of selective breeding.
- Evaluate the importance of selective breeding, and explore the ethical issues involved.

Learning outcomes

- Describe how we breed animals with features that we want in order to produce more organisms with the feature. [O1]
- Explain how farmers use the process of selective breeding – for example breeding dairy cows that produce large quantities of milk. [O2]
- Evaluate the importance of selective breeding in producing sufficient food for the world's population; explore the ethical issues involved in selective breeding – including the impact on biodiversity and the dangers of inbreeding. [O3]

Skills development

- Thinking scientifically: analyse data
- Working scientifically: interpret evidence
- Learner development: communicate effectively

Resources needed photographs of different dog breeds; Worksheet 3.1.5

Digital resources Quick starter; Interactive activity: Reorder the sentences to describe the process of selective breeding of cattle

Key vocabulary selective breeding

Teaching and learning

Engage

- Show photographs of different dog breeds – ask in what ways are the dogs similar (e.g. size of canine teeth) and in what ways are they different (e.g. coat colour). Ask the students to **identify** the features that owners value in the different dogs and the methods they use to breed the 'right' kind of puppy. [O1]
- **Think, pair, share** Working individually, ask the students to **identify** five features that might be important when breeding sheep. Then ask them to **compare** their list with a partner's and together **prioritise** the features. Take feedback and agree on the important features. [O1]

Challenge and develop

- Ask the students to read the 'Why do we use selective breeding?' section in the Student Book and to work with a partner to answer questions 1 and 2. Discuss the use of selective breeding in plants – for example how carrots were selectively bred from the wild carrot found in Iran. [O2]
- **Pair talk** Working in pairs, the students **brainstorm** answers to 'What other examples of selective breeding in plants and animals can you think of?'. [O1&2]
- The students work individually to complete task 1 of Worksheet 3.1.5. [O1]

  Higher-attaining students could **research** the origins and selective breeding of the Arabian saluki.

Explain

- Read and discuss 'The process' section in the Student Book. Ask them to work in pairs to **present** the example of selective breeding of dairy cattle as a diagram. [O2]

- The students work in pairs to answer questions 3 and 4 in the Student Book to **explain** the process of identifying desirable features and the process of selective breeding. Take feedback and discuss. [O2]

Consolidate and apply

- The students **apply their learning** by completing task 2 of Worksheet 3.1.5 concerning selective breeding in racing camels. [O1&2]
- **Group work** The students work in groups to **research** how new breeds of roses are bred. They **present** their explanation of the process as a flow diagram and **peer-assess** the work of others. [O2]
- **Envoys** Allocate each group a different example of selective breeding to **research** – to include Scottish fold cats; twisty/Kangaroo cat; broiler chickens; pigs and heart problems; sheep and breeding patterns. The envoys from each group visit other groups to **share** information. The class then **discuss** the ethics of selective breeding. [O1, 2&3]
- The students then work individually to complete task 3 of Worksheet 3.1.5. [O3]

Extend

- Ask students able to progress further to use reference books or the internet to research selective breeding for growing corn in hot climates. Ask them to **present** the information as a booklet that includes:
  - characteristics that are desirable for corn
  - how breeders select corn plants to breed
  - environmental factors that can affect the successful growth of corn plants (e.g. water and nutrients). [O1&2]

Plenary suggestions

What do we know? The students **work collaboratively** in a group of four to **design a short presentation** of their learning from this lesson. [O1, 2&3]

Answers to Student Book questions

1. taking organisms with desirable features and breeding them together, to produce more organisms with that feature
2. to produce plants with bigger crops or yields; to produce animals with more meat/milk/larger eggs etc.
3. *roses*: flower size; colour or scent; no prickles on stems; profuse flowers
   tomatoes: size; colour; flavour; number per stem etc.
4. select two plants with the desirable feature; transfer pollen from one to the stigma of the other; collect seeds; grow plants; repeat the process
5. farmers can produce larger crops more efficiently; but because organisms are so closely related, they are all at risk if a if a new disease emerges (e.g. bananas at risk of Panama disease)
6. as a minimum: discussion of benefits of new 'improved' breeds produced versus vulnerability to disease and the risk of deformities

Answers to Worksheet 3.1.5

1. a) i) Cotswold × Friesian ii) Cotswold × Derbyshire Gritstone

   b) any suitable examples and explanations: for example, alpacas with thick wool coat; alpaca wool is very soft.
2. a) lean and lightweight; small hump and tail

   b) select two camels with the desirable feature; let them mate; mate the offspring with another camel also bred from parents that have the desirable feature; keep following this process

   c) diet; amount and type of training; disease; health and veterinary care; weight of camel; conditions in which the animal is kept
3. a) selective breeding results in larger crops, animals that produce more meat etc., but because organisms are so closely related, all organisms are at risk if a new disease emerges; also danger of deformities (giving examples)

   b) The students should give valid arguments for their point of view with evidence. Broiler chickens produce large meaty chickens, but their body growth rate is much faster than their legs so they cannot stand; they may suffer heart problems and pain; if they cannot stand, they will not be able to reach food and water.

1.6 Finding out how organisms survive

Lesson overview

Learning objectives

- Describe how variation causes competition for resources, and drives natural selection.
- Explain the theories of Lamarck, Wallace and Darwin.
- Evaluate the importance of Darwin's work.

Learning outcomes

- Describe how differences between individuals (e.g. neck length of giraffes) causes competition for food, and describe how only those organisms that can get food will survive. [O1]
- Explain how the theories of Darwin and Wallace were similar (focusing on natural selection), and how Lamarck's theory was different (focusing on acquired characteristics). [O2]
- Evaluate the importance of Darwin's work in explaining how life has evolved. [O3]

Skills development

- Thinking scientifically: understand how theories develop
- Working scientifically: present evidence
- Learner development: act responsibly

Resources needed a piece of fabric the size of a tablecloth – preferably a different pattern for each group; 20 small card circles (made using a hole punch) in each of three different colours (60 in total) in a small bag; 20 extra small circles of each colour in separate bags; coloured pencils the same colour as the circles; graph paper; stopwatch; Worksheet 3.1.6; Practical sheet 3.1.6

Digital resources Quick starter; Interactive activity: Reorder the sentences to describe the process of evolution by natural selection; Slideshow: How life on earth evolved - A look at Charles Darwin's theory of evolution; Hangman: Key vocabulary game

Common misconceptions Natural selection is a random process. Evolutionary change in a species happens during a lifetime.

Key vocabulary natural selection, evolution

Teaching and learning

Engage

- Ask the students: 'What animals have been able to survive despite massive challenges? What characteristics would an animal need to survive for hundreds of millions of years?'. Take feedback and record the answers. [O1]

Challenge and develop

- **Group work** Working in groups of three, the students **brainstorm** why crocodiles and alligators have survived for over 65 million years. They **note down** all their ideas. [O1]
- Give the groups 10 minutes at the most to **research** prehistoric and modern crocodiles. Take feedback, asking the students to **give evidence** about how the animals have survived. [O1]

  Lower-attaining students could be provided with source material to use.

- Ask the students to read the 'What can cause a population to change?' section in the Student Book and to **discuss** questions 1 and 2 in their groups. They then complete task 1 of Worksheet 3.1.6. [O1]

Explain

- **Envoys** Home groups **research** either Lamarck's, Darwin's or Wallace's contribution to our understanding of evolution. The students should **write notes** about their findings. The envoys then visit other groups, who **teach** the envoy about their scientist – each envoy returns to their group to **share** information. Take feedback and tell the students to keep their work for later in the lesson. [O2]
- The students read the 'Which theory?' section in the Student Book and answer questions 3 and 4. [O2]
- Ask the students to work in groups of four to carry out the activity on Practical sheet 3.1.6 to **investigate** natural selection, and then answer the questions given. The groups will need a suitable space for their 'habitat' – such as two tables pushed together. [O1&2]

Consolidate and apply

- The students **write** a newspaper article about the three scientists and their ideas about evolution. Each group **peer-assesses** the work of another group and then you take feedback. [O2]
- Ask the students to complete tasks 2 and 3 of Worksheet 3.1.6. [O2&3]
- Working in groups the students **discuss** and **evaluate** the work of Charles Darwin and its continuing impact on science. [O3]

Extend

Ask students able to progress further:

- to **role-play** a court case – the prosecution has to try to prove Lamarck's theory to be correct, while the defence tries to prove Darwin's theory to be correct; the teacher acts as the judge and other students act as witnesses. [O1, 2&3]
- to **suggest** beneficial mutations that could arise in humans through natural selection [O1]
- to answer the question 'Why do organisms with better fitness generally leave more offspring than organisms that are less fit?'. [O1]

Plenary suggestions

The big ideas The students write down three ideas they have learned during the lesson. They **share** their facts in groups and **compile** a prioritised list of facts. Take feedback and find out which group(s) agreed. [O1, 2&3]

Answers to Student Book questions

1. the idea that organisms which are best suited to their environment survive and pass their features on to their offspring
2. it will be able to see over a longer distance and over obstacles (other animals, bushes etc.)
3. the idea that the more you use a feature the more developed it becomes; it is then passed on to offspring
4. no, your arm does not get longer; this refutes the theory of acquired characteristics
5. Darwin linked the idea of variation of a feature within a species with the survival advantage that the feature gave.
6. Darwin gave a scientific explanation for the variety of life; he argued that features change over generations to adapt to changing conditions, i.e. theory of evolution through natural section; scientists have developed his ideas further as their knowledge has increased.

Answers to Worksheet 3.1.6

1. a) Rabbits that run very fast have the best chance of not being caught by predators.

   b) *giraffe*: can reach more food; *porcupine*: quills prevent predation; *squirrel*: hibernates; *dolphin*: has tail/flippers; *polar bear*: coat provides camouflage and keeps it warm; *camel*: does not sink into sand; *lizard*: avoids predation
2. a) should include: a description of acquired characteristics (*Lamarck*); and of natural selection (*Wallace* and *Darwin*); natural selection is most convincing, because the evidence shows you cannot pass on features that develop in a lifetime b) A with iv; B with iii; C with i; D with ii
3. a) those that survive are better adapted (e.g. in terms of beak shape/structure) to eat the available food supply

   b) there was different food on each island; the finches' beaks had changed gradually to become better adapted

   c) no other birds lived there, so birds had never been a food source; large, predatory animals could not have reached the islands d) no other birds had flown to the islands because the islands were so far away from the mainland

1.7 Applying key ideas

The problem with dogs

Objectives

- Extract ideas about variation within a species from the text, including earlier topics.
- Apply ideas about variation to explain evidence.
- Apply ideas and information about selective breeding to propose the outcome of a process.

Outcome

- Clear and effective responses to questions, indicating understanding and the next steps in learning.

The purpose of this activity is to provide an opportunity to see how successfully students are grasping the key ideas so far. It isn't designed to be used as a formal test – it might be that the students work on the questions collectively. It does provide an opportunity for you to look at written work, engage students in discussions and form ideas about progress being made.

The tasks are progressive. Lower-attaining students should be able to tackle the first two or three tasks, and higher-attaining students will find the later ones more challenging.

Resources needed coloured pencils; highlighter pens; a picture of four species of mammal (e.g. brown bear, polar bear, koala and panda); Worksheet 3.1.7

Teaching and learning

Engage

- Challenge the students to **reflect back** over the last few lessons and **identify** some of the key ideas they have learned about. Draw out responses about these ideas and say that it is important as learners that they develop a sense of their own progress.
- **Think, pair, share** The students work in pairs to **draw a concept map** of the key ideas identified. They swap maps with another pair and use this for a 'back-to-back' activity. One student **describes** the other pair's concept map to their partner, who has to **replicate** this without seeing it. Have coloured pencils ready on the desks. Take feedback – how well did they do?

Challenge and develop

- Give each pair a key idea from the topic and a sheet of flip-chart paper. Ask them to **summarise** everything they can remember about the idea in 30 seconds. Each pair passes their work on to the next pair – repeat the process three or four times. Return the work to the original pairs and give them time to read it through. Each pair then **presents** their ideas to the class. Encourage the students to **ask questions** and **discuss** the ideas.
- Show the students a picture of four species of bear. In pairs they **discuss** the questions 'How are these animals similar? How are they different? Why is biodiversity important?'.

Explain

Ask the students to read 'The problem with dogs' in the Student Book. Depending on their ability to assimilate text, you may need to adopt strategies such as:

- Have the students working in pairs using copies of the text (provided as Worksheet 3.1.7) and highlighters. One student highlights key ideas and the other (using a different colour) any words they do not understand. Take feedback and ask other students if they can **explain** the difficult words identified.

Consolidate and apply

- Now ask the students to attempt the tasks in the Student Book – they might do this individually or collaboratively. In each case, encourage them to **identify** and **record** their ideas.
- Ask the students to **present** their responses to various tasks, either orally or by displaying their work.

Extend

- For higher-attaining students, task 6 in the Student Book gives an opportunity for extension work.

Plenary

How did you do? Ask the students for their responses to questions such as:

- How well were you able to approach those tasks?
- Can you improve on some of these responses by working together as a class?
- Are there some ideas that you need to return to at a later stage? (If there are, make a note.)

Provide the class with some overall feedback, indicating ideas that they have grasped effectively and those that may need developing further.

Likely responses and next steps in learning

1. This task gives the students an opportunity to recall learning experiences, or to refer to text in the book, about variation between individuals within a species. They should be able to explain that dogs vary in terms of: height; leg length; tail shape and length; size and shape of ears; colour of fur; length of fur.
2. This task involves recalling first a feature that changes gradually over a range of values has continuous variation, and then that some features have only a limited number of possible values – there is no range and individuals have one type of the feature or another – this is discontinuous variation. Examples of continuous variation are height; weight; leg length; tail length. Examples of discontinuous variation are shape of face; shape of ears; shape of tail; colour of fur.
3. This task involves recalling and explaining how selective breeding can produce different breeds of organisms. The students should recall that desirable traits can be identified, and that organisms with these traits are bred together. This continues until a new breed has resulted that has all the desirable traits. The students may need prompting to remember that the traits must be bred for over a number of generations.
4. This task involves the students explaining what they have learned about the importance of genetic variation. Genetic variation is needed for the process of natural selection (through survival advantage) to produce evolutionary change whereby a species changes gradually and survives. Selective breeding has limited survival advantage in many dog breeds. There is a very limited range of the features that are seen as desirable in pedigree dogs. Any change in their environment could eradicate the breed.
5. The students need to apply their learning and evaluate the issues of selective breeding in a new context in this task. Some dog breeds are still working dogs – for example: Border collies are used as sheepdogs and in search and rescue operations; Labrador retrievers are used as gundogs in hunting activities. It is important to keep the desirable features in the breeds as well-developed as possible, while ensuring that inbreeding does not become an issue because this could result in inherited problems. There is a new direction in dog-breeding: two breeds are bred to produce crosses – for example, labradoodles are labradors crossed with poodles. This is introducing inherited variation.
6. This task involves the students thinking across the key ideas that they have explored and selecting the relevant ideas that apply to dog-breeding. Possible ideas include the following:
    - Inherited variation means that different breeds have different features that make them suitable to undertake different roles.
    - Selecting desirable features for working dogs ensures that they are able to do a job to a high standard – for example guard dogs, guide dogs for the blind, dogs used to detect drugs.
    - Selective breeding of features that cause discomfort, pain or poor quality of life is not ethical.
    - Using close relatives for breeding, which increases the risk of medical (or behavioural) problems, is not ethical.
    - Breeding pedigree dogs is not wrong, but it is important to ensure that unrelated dogs are bred in order to try to maintain the variation.
    - Selective breeding of cross-breeds is increasing inherited variation.

1.8 Understanding why siblings are different

Lesson overview

Learning objectives

- Identify inherited features in plants and animals that vary between offspring.
- Explain how inherited differences arise by genetic material from both parents combining.
- Describe how identical twins occur and analyse data about their features.

Learning outcomes

- Identify inherited differences between members of the same family. [O1]
- Explain how inherited differences arise in members of the same family. [O2]
- Predict and research how identical twins occur and analyse evidence about their features. [O3]

Skills development

- Thinking scientifically: analyse data
- Working scientifically: record evidence
- Learner development: respect others

Resources needed small mirrors; photos of a black Labrador and a Dalmatian; five tomatoes of three different types (e.g. piccolino, santini and beef tomatoes); at least two very different flowers, vegetables or fruits; Worksheet 3.1.8; Practical sheet 3.1.8

Digital resources Quick starter; Video; Interactive activity: Drag the statements about twins to the correct group

Common misconceptions Identical twins are actually always identical.

Key vocabulary inherit, trait, chromosome, identical twins, conjoined twins

Teaching and learning

Engage

- Show the students five tomatoes of two/three different types. Ask them to **identify** features that tomatoes inherit from their parents. Record the students' responses. [O1]
- Show photos of a black Labrador and a Dalmatian. In groups, the students **predict** and **draw** the features of puppies produced by breeding from these parents. Discuss their ideas – what could happen to the puppy to alter how it looks? [O1]

Challenge and develop

- The students work individually to complete task 1 of Worksheet 3.1.8. They **decide** whether certain features are inherited, may be inherited or are not inherited. [O1]
- The students work in pairs to **carry out** the task on Practical sheet 3.1.8. They **record** their individual traits and then **compile** class data to answer the questions.
- Look at other organisms, such as garden flowers, fruit or vegetables, and ask the students to **identify** characteristics that might be inherited. [O1]

Explain

- The students read the 'Inherited features' section in the Student Book, **discuss** questions 1 and 2 with a partner and **agree on** answers. [O1]
- **Think, pair, share** Ask the students to **consider** part a) in task 2 of Worksheet 3.1.8. Then ask 'Why do some members of the family share common features that are not shown by others?'. Establish that this may be because inherited characteristics were not passed from parent to each offspring. [O1&2]

- Ask the students to **suggest** how offspring inherit characteristics from their parents – remind them of their work on reproduction and cells. Establish that during fertilisation, animal and plant cells pass on 'information' in the cell nuclei from one generation to another. [O2]

  Higher-attaining students may be able to identify that information is passed on in genes/chromosomes.
- Ask the students, working in pairs, to read the 'Why are siblings different?' section in the Student Book. They then **draw diagrams** to clarify what happens to genetic information before and after fertilisation. [O2]

Consolidate and apply

- Working in groups, the students **draw a sequence of diagrams** on a poster illustrating the process of sex-cell formation and fertilisation. Using annotations, they show how genetic information is transferred and how sperm and egg cells are adapted. [O2]
- The students then work in pairs to complete parts b) and c) of task 2 of Worksheet 3.1.8. [O2]
- The students work in groups to **predict** and **research** how identical twins are formed. Ask the groups to **analyse** the features of the girls in Figure 3.1.8d in the Student Book and answer the question in the caption. Take feedback and establish that identical twins are not always exactly the same. [O3]
- Working in pairs, the students read the 'Identical twins' section in the Student Book and answer questions 5 and 6. They can then complete task 3 of Worksheet 3.1.8. [O3]

Extend

- Ask students able to progress further to **research** the occurrence of identical twins in other animals. Take feedback and discuss the information. [O3]

Plenary suggestions

Learning triangle The students draw a large triangle with a smaller inverted triangle inside. In the three outer triangles they write something they've seen; something they've done; something they've discussed. In the central triangle they write something they've learned. [O1, 2&3]

Answers to Student Book questions

1. it is a feature that is passed on from parents to offspring
2. any ten inherited traits, such as eye colour, hair colour, shape of face, dimples, curly hair, etc.
3. Each sex cell contains a random mix of genetic information, passed on from four grandparents.
4. in the chromosomes
5. One egg is fertilised by one sperm and then splits into two.
6. any suitable ideas, e.g. different diet; ill health; scarring; dying hair colour; different hair style; tattoos; plastic surgery

Answers to Worksheet 3.1.8

1. a) i) is inherited ii) is not inherited iii) may be inherited iv) may be inherited v) is inherited vi) is not inherited vii) is inherited viii) is not inherited ix) is inherited x) may be inherited
2. a) i) shape of head, body, mouth, eyes, hands and feet; curly, straight or no hair; colour of eyes, head and body

   ii) *eye shape*: inherited genetic information from his mother; *curly hair*: inherited genetic information from his father

   iii) hair – father; head shape – father; body shape – father; head/body colour – grandmother; eye shape and colour – father; mouth shape and colour – father; hands and feet – father

   iv) Baby 2 has inherited different genetic information/chromosomes from the mother and the father.

   b) The ovule and pollen contain a random mix of genetic information from the two parent plants.

   c) chromosomes
3. *Identical twins* form when one egg is fertilised by one sperm and the egg then divides into two halves. The twins have the same genetic information and look the same.

   Conjoined twins are identical twins that are joined together in some way; they are formed in the same way, but the fertilised egg does not completely split into two.

   b) After the fertilised egg splits in two, genetic information in the chromosomes can sometimes change when cells continue to divide; there may also be the effects of environmental factors (as mentioned in Student Book question 6 above) – for example diet can affect height and/or weight.

1.9 Looking inside a cell's nucleus

Lesson overview

Learning objectives

- Identify that the nucleus contains chromosomes, which carry inherited genetic information.
- Explain that chromosomes are made of genes containing DNA, and describe the structure of DNA.
- Assess the work of Watson, Crick, Wilkins and Franklin on DNA structure.

Learning outcomes

- Identify that cells have nuclei containing chromosomes that carry genetic information. [O1]
- Explain and model the structure of DNA; explain the link between chromosomes, genes and DNA. [O2]
- Analyse the role of different scientists in the discovery of the structure of DNA. [O3]

Skills development

- Thinking scientifically: understand how theories develop
- Working scientifically: develop explanations
- Learner development: act responsibly

Resources needed liquorice sticks; coloured marshmallows (in four different colours); toothpicks; paperclips; masking tape; self-closing plastic bags; chemical modelling kits or materials such as card, straws, string, modelling clay and polystyrene; Worksheet 3.1.9a; Worksheet 3.1.9b; Practical sheet 3.1.9 (preferably printed in colour)

Digital resources Quick starter; Interactive activity: Complete the sentences about DNA

Key vocabulary DNA, gene, base, double helix

Teaching and learning

Engage

- **Think, pair, share** Ask the students 'What do a banana, chimpanzee and a human have in common?'. (We all share much of the same DNA.) [O1]

Challenge and develop

- **Small groups** Ask the students to read the 'Inside the nucleus' section in the Student Book and then to **summarise** the key information in no more than 30 words. [O1]
- **Three stars and a wish** Ask student pairs to **draw** a simple diagram to show the link between a cell, a nucleus, a chromosome, a gene and DNA. Each pair **peer-assesses** the work of another pair, identifying three 'stars' (things they have done well, learned, understood or enjoyed) and one 'wish' (something they could improve, need to complete or need help with). The students then amend their own work if needed. [O1]

  The diagrams need to be correct for the next activity.
- **Back-to-back** Pairs exchange diagrams – the two students in each pair sit back-to-back. Student A **describes** the diagram to student B (who cannot see it). Student B **asks questions** to clarify their understanding and **draws** the diagram as accurately as possible. After a set time, the students **compare** the diagram and the drawing. [O1]

Explain

- The students **work independently** to answer questions 1 and 2 in the Student Book and complete task 1 of Worksheet 3.1.9a. [O1]

- Ask the students to read the 'DNA' section in the Student Book and look at Figure 3.1.9b, which helps to envisage the structure of a DNA molecule. The students work in pairs to answer questions 4 and 5. [O2]
- **Group work** The students work in groups of four to **make models** of DNA using Practical sheet 3.1.9. Each group will need the supplies listed in the Practical sheet – you may find it helpful to prepare supplies in advance for each group, putting everything they need into a plastic bag. To make the marshmallows easier for students to pierce with a toothpick, let them dry out slightly by snipping holes in the marshmallows bag(s) on the day before the activity.

  When the students have completed the activity, assess their understanding of the structure of DNA by examining whether or not they have built and labelled their models correctly – don't forget to check the base pairing. [O2]
- The students then complete task 2 of Worksheet 3.1.9a. [O2]
- Ask the students to read the 'Finding the structure' section in the Student Book and to answer questions 6 and 7. They then complete task 3 of Worksheet 3.1.9a. [O3]

Consolidate and apply

- The students work in pairs to **plan** a talk outlining the structures involved in heredity (cell, nucleus, chromosome, DNA, gene). They must **use their models** to illustrate their talks. The students **present** their talks to one other group, and each group **peer-assesses** the work of the other group. They can use Worksheet 3.1.9b to **make notes** on their assessment. Take feedback. [O2]

Extend

- Ask students able to progress further to use a modelling kit to **construct an accurate model** of the structure of DNA. Alternatively they could construct a model using art supplies such as card, straws, string, modelling clay and polystyrene. [O2]

Plenary suggestions

The big ideas Ask the students to write down three things they have learned about DNA. Then ask them to **share** their facts in groups and to **compile** a master list, with the most important at the top. Ask for ideas to be **presented** and find out which other group(s) agreed. [O1, 2&3]

Answers to Student Book questions

1. A *gene* is a region of DNA; *chromosomes* are made up of lots of genes.
2. Genes control the development of different characteristics by giving instructions to the cell.
3. 46 (accept 23 pairs)
4. two strands in a helix, with 'rungs' of base pairs
5. A–T; C–G
6. Franklin demonstrated the helical nature of DNA using X-ray diffraction.
7. Wilkins and Franklin discovered the helical nature; Chargaff discovered the base pair matching; these scientists shared their findings with Watson and Crick.

Answers to Worksheet 3.1.9a

1. a) i) in the cell ii) contains chromosomes iii) made of DNA; divided into genes iv) passed from generation to generation; to determine features

   b) suitable diagram
2. correct diagram illustrating double helix, with correct base pair 'rungs'
3. a) Answers depend on the format chosen, but research should build on information in the Student Book and exemplify the work of each scientist.

   c) i) the idea that scientists worked in competition, but also collaborated to build on each other's work

   ii) ideas to include: Watson and Crick had no expertise in X-ray crystallography, and needed the research both of Franklin and Wilkins on the helical structure, and of Chargaff on the pairings of the bases, to build the accurate model

1.10 Learning about DNA

Lesson overview

Learning objectives

- Identify that all plant and animal cells contain DNA.
- Explain why it is important for scientists to be able to extract DNA from cells.
- Analyse and evaluate the use of extracted DNA.

Learning outcomes

- Identify that DNA is in the food we eat. [O1]
- Explain why it is important for scientists to extract DNA from organisms. [O2]
- Explore the ethics and evaluate the use of extracted DNA. [O3]

Skills development

- Thinking scientifically: analyse data
- Working scientifically: develop explanations
- Learner development: communicate effectively

Resources needed strawberries; prepared slides of root tip squashes; microscopes; equipment and materials as detailed in the Technician's notes; Worksheet 3.1.10a; Worksheet 3.1.10b (copied onto card); Practical sheet 3.1.10; Technician's notes 3.1.10

Digital resources Quick starter; Slideshow: Working with DNA - A look at DNA extraction and its use in forensics; Interactive activity: Drag the statements about DNA data to the correct group - ethical issue, scientific issue, or both

Key vocabulary biotechnology, forensic science, genetic disorders, DNA profile

Teaching and learning

Engage

- **Think, pair, share** Ask the students: 'How do you think scientists get DNA out of cells? What difficulties could there be in doing this?' Record their ideas on the board to see how well these questions are answered as the students do the practical later in this lesson. [O2]
- **Pairs to fours** Ask the students to **discuss** the question 'Have you ever eaten DNA?'. Take feedback and discuss their answers to elicit that the food we eat was once a living plant or animal, and that all living things have DNA. Remind the students about the 2013 news story in which horsemeat was found in beef burgers and other foods. Ask: 'How did scientists discover it was horsemeat and not beef?'. Elicit that the horsemeat was detected by DNA tests. [O1]

  Higher-attaining students may make the link immediately while other students will need prompting.

Challenge and develop

- **Pair talk** Ask the students to read the 'DNA everywhere!' section in the Student Book and then answer questions 1 and 2 in pairs. Challenge the students by asking how we know that eating DNA does not affect us adversely. Ask them to look at Figure 3.1.10a and **discuss** the question in the caption. Ask them to **imagine** what might happen to the sheep if they were affected by the grass DNA they ate (they would probably turn green!) [O1]
- Ask the students if they have ever seen DNA, a gene or a chromosome. The students then **observe** prepared root tip squashes under the microscope and **draw** what they see (cells at various stages of cell division). [O2]

Explain

- Working in pairs, the students **research** and **write** a newspaper article about the use of DNA in the horsemeat scandal across Europe in 2013 using task 1 of Worksheet 3.1.10a. [O1&2]
- The students read the 'Extracting DNA' section in the Student Book and answer questions 3 and 4. [O2]
- Demonstrate the extraction of DNA from strawberries – see Technician's notes 3.1.10. Talk through what you are doing and ask the students to **observe** closely. Point out the changes at the interface as the DNA strands become visible. The students can answer the questions provided on Practical sheet 3.1.10. [O2]

Consolidate and apply

- Working in pairs, the students **sequence the steps** for the extraction method using the cards from Worksheet 3.1.10b. [O2]
- The students read the 'How should we use DNA?' section in the Student Book and work in groups to **discuss** their answers to questions 5 and 6. [O3]

Extend

- Ask students able to progress further to **research** the use of extracted DNA, working in small groups. The students **evaluate** the use of extracted DNA and **explore the ethics** around its use. They **present** their work in a slideshow of three slides at most using task 3 of Worksheet 3.1.10a. [O3]

Plenary suggestions

Summarise Ask students to work in pairs to **summarise** the lesson in no more than 30 words by referring to the questions in the first *Engage* activity. Take feedback and the students **peer-assess** the statements. [O1, 2&3]

Two stars and a wish Alternatively students could work individually to identify two 'stars' (things they have done well, learned, understood or enjoyed) and one 'wish' (something they could improve, need to complete or need help with). [O1, 2&3]

Answers to Student Book questions

1. Our food comes from living things; all living things have DNA.
2. Most DNA denatures and breaks down; some passes out in faeces.
3. it is used to detect genetic disorders; to make beneficial products such as insulin, antibiotics and hormones
4. blood; skin; saliva
5. innocent people can be identified; genetic diseases detected; the likelihood of contracting a disease can be identified (e.g. cancer); it can be used to determine whether people are related or not
6. *Advantages*: a national DNA database helps to identify offenders by matching DNA from a crime scene to known criminals; it also permits other identity checks.

   Disadvantages: possible errors in interpreting DNA; being over-confident about its accuracy; security issues in making sure data are not lost or stolen, or database hacked; issues around including innocent people on the database; some argue it goes against an individual's civil liberties (invasion of privacy).

Answers to Worksheet 3.1.10a

1. The article should include the fact that the Food Standards Agency monitors food products for 'foreign' DNA, and was able to determine where products had been made using DNA testing.
2. Cells are ground up; a salt solution and a cleaning agent are added to break down fat and proteins in cell membranes; ethanol is added to solidify any DNA.
3. aspects to cover include: use of hairs or saliva left on clothing worn during a robbery; semen located at a rape scene; blood collected from an assault suspect's clothing; 'wearer DNA' on clothing; traces of assailant's skin or blood under a victim's fingernails; use of mitochondrial DNA to identify maternity; Y-chromosome anlysis for paternity; identification of missing persons or unknown remains; *advantages* include: ability to identify perpretrators of crimes many years after the event; an increased chance of identifying perpetrators accurately; *limitations* include: trace evidence could lead to incorrect identifications; complex testing and storage could lead to errors

Answers to Worksheet 3.1.10b

The correct order is: D, C, F, B, A, E.

Answers to Practical sheet 3.1.10

1. to break open the strong cell walls 2. to dissolve the layers of the cell membrane and cell organelles 3. to break up the protein chains that bind around the nucleic acids 4. alcohol 5. no; it will have clumps of protein attached to it
6. DNA is colourless, but the coloured protein clumped with it makes it visible.

1.11 Exploring human chromosomes

Lesson overview

Learning objectives

- Identify that, at fertilisation, one chromosome in each pair comes from each parent.
- Explain how fertilisation results in each new individual being genetically unique.
- Explain how some genetic disorders arise.

Learning outcomes

- Recall that during fertilisation one chromosome in each pair comes from each of the parents. [O1]
- Explain that the fusion of male and female sex-cell nuclei (in both animals and plants) produces a new individual that is genetically unique. [O2]
- Explain the impact of slight 'changes' to the DNA passed on from the parents. [O3]

Skills development

- Thinking scientifically: ask questions
- Working scientifically: develop explanations
- Learner development: collaborate effectively

Resources needed string; scissors; sticky tape; different coloured pipe cleaners; coloured pens; Worksheet 3.1.11a (the second page printed in colour on card); Worksheet 3.1.11b; Practical sheet 3.1.11

Digital resources Quick starter; Interactive activity: Reorder the sentences to describe the process of sexual reproduction in humans; Interactive activity: Drag the symptoms to the correct genetic disorder - Down's syndrome or cystic fibrosis; Slideshow: Chromosomal disorders - Explanation and examples

Key vocabulary karyotype, trisomy

Teaching and learning

Engage

- **Think, pair, share** Ask the students the question 'How is sexual reproduction similar to and different from cell division?'. Take feedback and discuss their ideas. Elicit that the sperm first fertilises the egg, and then the cell divides resulting in a new organism. [O1]

  Lower-attaining students may need some prompting through questioning.

- **Pairs to fours to eights** Working in pairs, ask the students to **discuss** this question: 'Why is every human being different?'. The pairs join to **share** ideas in groups of four, and then eight. Take feedback and discuss their ideas. [O1]

Challenge and develop

- **Pair talk** The students work in pairs to read the 'Making a new individual' section in the Student Book. They **discuss** why gametes have half the normal chromosome number and then answer questions 1 and 2. [O1]
- The pairs **investigate** how chromosomes are passed down to offspring by **using a simple model**, as instructed on Practical sheet 3.1.11. They can use a variety of materials, or even computer animations, to show the behaviour of chromosomes and what happens in gamete formation and fertilisation. [O1]
- The students then **discuss** and answer the questions in task 1 of Worksheet 3.1.11a. [O1]

Explain

- **Genes lucky dip** The students now **carry out** the task on Worksheet 3.1.11b to **design** a dog with genetic traits based on randomly chosen DNA codes. This activity can be done individually or in pairs. Give each student or pair an envelope labelled 'Dog DNA'. This should contain 32 strips of paper (eight copies of each the four DNA codes on page 2 of Worksheet 3.1.11b). The students **select** codes randomly from the envelope and **match** these to the 'Dog traits key', thereby identifying their dog's particular genetic features. Explain the activity and then circulate to support the students as they work through the task. When all the students have completed their dog's DNA code and **drawn** their dog, display the results. Emphasise that all the dogs are different because of the random selection of genes. Ask the students to **evaluate** this model of how traits are inherited – 'What was good about the model and how could it have been improved?'. [O2]

- **Hot potato activity** In pairs, the students read the 'Different and unique' section in the Student Book. Each pair has a different coloured pen. They have 30 seconds to start **writing** an answer to question 3. They then pass their part-written answers on to the next pair, who **amend** and **add to** the answer. Repeat this until a complete answer is written (5 minutes maximum). Take feedback and develop a class answer. [O2]
- **Pair work** Ask the students to answer question 4 in the Student Book and then **work together** on task 2 of Worksheet 3.1.11a, using the chromosome cards from page 2 of the worksheet. [O2]

Consolidate and apply

- The students read the 'When things go wrong' section in the Student Book and then **discuss** what they have read with a partner and **agree on** answers to questions 5 and 6. [O3]
- **Group work** The students work in groups to **compare** the karyotype of Down's, Edward's, Patau's, Kleinfelter's and Triple X syndromes to the normal karyotype in the Student Book, and **list** the differences. [O2&3]
- **Envoys** Each home group **researches** one of the following: Down's, Edward's, Patau, Kleinfelter's and Triple X syndromes (in task 3 of Worksheet 3.1.11a). Each group's envoy visits another group, who **teach** the envoy about the syndromes. The envoys then return to their groups and **design** a scientific poster about chromosome defects. [O3]

Extend

Ask students able to progress further:

- to **design** a webpage to display their research on chromosomal defects [O2]
- to **research** the discovery and identification of a gene responsible for a particular genetic disorder. [O3]

Plenary suggestions

Information tweet Ask the students to **write a tweet** to a friend about two interesting things they have learned in the lesson. [O1, 2&3]

Exit slips Alternatively students work individually to identify one key thing they have learned in the lesson. They write this on a sticky note that they stick to the board as they leave the room. [O1, 2&3]

Answers to Student Book questions

1. the nucleus
2. so that the fertilised egg has the normal full set of chromosomes found in body cells
3. the chromosome they inherit in every pair from each parent is random
4. a set of chromosomes
5. they should include: extra or fewer chromosomes; changes to a single gene or group of genes (deletions or duplications); inheriting a form of the gene that results in a particular condition
6. the presence of three chromosomes instead of a normal pair, e.g. trisomy 21 (Down's syndrome)

Answers to Worksheet 3.1.11a

1. a) there are 23 in each egg cell and sperm cell; 46 in a fertilised egg
   b) so that the fertilised egg has the normal full set of chromosomes found in body cells

Answers to Worksheet 3.1.11b

1. It is very likely that the dogs are all unique.
2. their features are determined by a unique set of genetic codes

1.12 Understanding cloning

Lesson overview

Learning objectives

- Define cloning and describe some natural cloning processes.
- Explain how organisms may be artificially cloned.
- Explore ethical issues around artificial cloning.
- Compare and contrast asexual and sexual reproduction.

Learning outcomes

- Define cloning as making identical copies of the single parent and describe examples – such as runners, bulbs, corms. [O1]
- Explain that genetic cloning involves removing the nucleus of an egg cell from one organism and replacing it with the nucleus of a normal cell of the organism to be cloned. [O2]
- Explore the advantages and disadvantages of cloning; compare asexual reproduction and sexual reproduction in terms of number of parents, numbers of offspring produced and the genetic variation in offspring. [O3]

Skills development

- Thinking scientifically: evaluate risks
- Working scientifically: develop explanations
- Learner development: communicate effectively

Resources needed materials and equipment as detailed in the Technician's notes; Worksheet 3.1.12; Practical sheet 3.1.12; Technician's notes 3.1.12

Digital resources Quick starter; Interactive activity: Reorder the sentences to describe how to clone a spider plant; Interactive activity: Reorder the sentences to describe how Dolly the sheep was cloned; Video

Key vocabulary clone, asexual reproduction, nuclear transfer

Teaching and learning

Engage

- **Think, pair, share** Working in pairs, ask the students to **brainstorm** all the different ways that plants and animals can reproduce. Take feedback and record all the asexual reproduction methods. [O1]
- **Pairs to fours** Ask the students, working in pairs and then in fours, to **discuss** these questions: 'What is a clone? Would you like to be cloned? If you had a clone of yourself, what would you use it for?'. [O1]

Challenge and develop

- Back in their pairs, the students read and **discuss** the 'Natural clones' section in the Student Book, answer questions 1 and 2, and then complete task 1 on Worksheet 3.1.12. [O1]
- **Envoys** In groups of four, the students **research** one of the following methods of cloning – via bulbs, corms, rhizomes, tubers, budding or parthenogenesis. Each group's envoy visits a different group to **teach** them about the method researched. The envoys return to their original group and **share the learning** to clarify the processes. [O1]
- Give the students samples of bulbs, corms, rhizomes, tubers etc. on white tiles or plates and ask them to **observe** them and **draw** annotated diagrams to show the asexual reproductive process in these. [O1]
- The students then use Practical sheet 3.1.12 to **investigate** the effect of growing cuttings using hormone rooting powder or the effect of growing cuttings placed either in plastic bags or not. The whole class can complete just one of these investigations, or groups could be allocated one of the options. [O1]

Explain

- Ask the students to work in groups to **draw** a comic-strip style illustration of how to take plant cuttings. [O1]
- In pairs, the students read the 'Artificial clones' section in the Student Book and answer questions 3 and 4. [O2]
- Still in their pairs, they **research** and **write** a short explanation of tissue culture. In the same pairs, the students complete task 2 of Worksheet 3.1.12. [O2]

Consolidate and apply

- **Small groups** The students read the 'Cloning: good or bad?' section in the Student Book and **discuss** the advantages and disadvantages of cloning (natural and artificial). They then answer questions 5 and 6 and complete task 3 of Worksheet 3.1.12. [O3]
- Ask the students to **summarise** their learning in no more than 30 words, using pictures and diagrams to **present** information. They **peer-assess** the work of other groups. [O1, 2&3]

Extend

Ask students able to progress further to work in pairs to:

- **discuss** and **explain** why cloning extinct animals is problematic [O1, 2&3]
- **discuss reasons** why a person might want to clone a human. [O3]

Plenary suggestions

Mapping from memory Use a good summary of learning from the *Consolidate and apply* section for students to map from memory. Stick the summary to your desk. Students work in groups of four, and take turns to **observe** the summary for 30 seconds and then **replicate** what they have seen in 30 seconds, before the next person looks at the summary. They can talk and ask questions within the group as they draw. [O1, 2&3]

Answers to Student Book questions

1. it is an organism that is identical to its parent 2. they all contain the same genetic material 3. Natural cloning occurs in nature (e.g. spider plants); artificial cloning only happens when humans intervene.
4. An egg cell was taken from sheep A, and its nucleus removed. A nucleus was removed from an udder cell of sheep B, and placed into the egg cell from sheep A. This egg cell was put into sheep A (or another sheep) to grow, and became Dolly. 5. because clones have genetic material from only one parent
6. *Advantage*: e.g. can create an exact copy of an organism with desireable features. *Disadvantage*: e.g. organisms have a shorter life span.

Answers to Worksheet 3.1.12

1. a) Cloning is a method of reproduction where all of the genetic material comes from one parent.
   b) they reproduce from root fragments; so to get rid of the weed all of the root system has to be removed
   c) *plants*: bulbs, tubers, rhizomes; *animals*: budding, parthenogenesis
2. the poster should include: taking an egg cell; removing its nucleus; replacing it with the nucleus from a normal body cell of the animal they want to clone; implanting the egg back into the first animal (the one it came from)

3.	**Sexual reproduction**	**Cloning**
Relative complexity of organism (including size)	Complex organisms tend to reproduce sexually.	Simple organisms tend to reproduce asexually.
Number of parents who contribute genetic information to the offspring	Two parents each contribute half of the genetic information. All offspring are unique.	One parent contributes all of the genetic information. Offspring are all identical.
Reproductive mechanism	Fusing of a male and a female gamete from two parents.	No fusion of gametes involved.
Genetic variation in offspring	Two parents result in genetic variation in offspring – this gives a survival advantage if the environment changes. Variation allows natural selection for adaptation development.	There is no variation, so if a change in the environment occurs no organism has a survival advantage. Well-adapted traits are passed on, which gives offspring a survival advantage.
Relative number of offspring produced	Time and energy spent searching for a mate decreases the number of offspring.	Many more offspring are possible.

1.13 Explaining extinction

Lesson overview

Learning objectives

- Identify changes that can cause a species to become extinct.
- Explain the use of gene banks to preserve hereditary material before a species becomes extinct.
- Analyse and evaluate theories of what caused the extinction of the dinosaurs.

Learning outcomes

- Identify possible causes of why some species became extinct – for example the dodo and woolly mammoth. [O1]
- Explain the use of gene banks for preserving endangered species. [O2]
- Analyse and evaluate the possible causes of the extinction of dinosaurs. [O3]

Skills development

- Thinking scientifically: ask questions
- Working scientifically: evaluate data
- Learner development: collaborate effectively

Resources needed Worksheet 3.1.13

Digital resources Quick starter; Hangman: Key vocabulary game

Key vocabulary extinction, endangered species, gene bank, mass extinction

Teaching and learning

Engage

- **Think, pair, share** Working in pairs, the students **discuss** the question 'What do dinosaurs, dodos, sabre-toothed cats and the woolly mammoth all have in common?'. Working in the same pairs, the students **agree on** a definition of the term 'extinction'. They **brainstorm** ideas and **write an explanation** of what they think caused the extinction of these animals. [O1]

Challenge and develop

- Ask the students to read the 'Causes of extinction' section in the Student Book. As a whole class discuss question 1 and develop a class explanation of when a species is officially classified as being extinct. [O1]
- **Envoys** Divide the class into groups of four. Each group **researches** one of the following – great auk, Irish elk, dodo, woolly mammoth, sabre-toothed tiger, using task 1 of Worksheet 3.1.13. Each group's envoy then visits another group and **teaches** them about their animal. The envoys return to their own group and **share learning** and works together to **write fact sheets** for each animal. [O1]
- Each group then **designs** a poster **explaining**, for the animal they researched, its habitat and environment, its food and possible causes of how it became extinct. [O1]

Explain

- The students work in groups to **discuss** question 2 in the Student Book and **draw** a cartoon strip summarising the possible causes of extinction. [O1]
- **Pair work** The students read the 'Preventing more extinctions' section in the Student Book and then **research** the use of gene banks and seed banks to preserve the DNA of plant and animal species. They then **discuss** and answer question 4 in the Student Book. [O2]

Consolidate and apply

- The students work in pairs to complete task 2 of Worksheet 3.1.13. [O2]

  Higher-attaining students should **evaluate** the use of gene banks and cloning in preserving species.

- Introduce Celia, the Pyrenean ibex, and ask student groups to **research** and **write a fact sheet** about how she was brought back to life after extinction. [O2]
- **Pair talk/Pairs to fours** The students read the 'Mass extinctions' section in the Student Book, answer questions 5 and 6, and **compare** answers. [O3]

  Some students may need prompting to think about variation and adaptation in question 6.

- The groups then **research** the possible causes of dinosaur extinction and complete task 3 of Worksheet 3.1.13. [O3]

Extend

- Ask students able to progress further to **discuss** the questions: 'Is prevention better than cure? Can we change the habits of humans to prevent extinction of endangered species?'. Take feedback and discuss the students' ideas. [O1, 2&3]

Plenary suggestions

What have we learned? Give the students one minute, in groups, to **brainstorm** everything they have learned this lesson. Take feedback and ask questions for areas they missed. [O1, 2&3]

Answers to Student Book questions

1. when there are no more individuals of the species alive in the world 2. *natural causes*: climate change; changes in sea level; asteroid impacts; disease *caused by humans*: hunting; habitat destruction; overexploitation 3. it is a store of DNA from plants and animals 4.Plant genes can be stored in the form of seeds.
5. effects include: loss of biodiversity; predators will take longer to recover; possible environmental change
6. variation within the species meant that some were better adapted to the changes

Answers to Worksheet 3.1.13

1.

Animal	**Kind of environment**	**Main source of food**	**Features that were a disadvantage**	**Environment changes**	**Reason for changes**
Great auk	Rocky, isolated islands	Fish	Flightless; not agile on land	Hunted for food	Sailors arrived on the islands
Dodo	Island of Mauritius	Fruits	Flightless; slow mover	Hunted for food by sailors and invasive species	Habitat destruction Eggs and adults eaten
Woolly mammoth	Very cold tundra	Grass	Very slow mover	Ran out of grass Hunted by humans	Global warming Used for food
Irish elk	Grasslands and hills in Europe, Asia and Africa	Grass, leaves etc.	Huge antlers that slowed it down	Unable to adapt to cold conditions	Ice age
Sabre-toothed tiger	Grasslands, shrubby areas and pine forests (worldwide)	Large animals e.g. mammoths	Overspecialised teeth could not adapt to smaller prey	Food loss	Large prey became extinct

2. this should include: genetic material from plants and animals (including plant seeds) are stored; then when needed, they are either grown (plant seeds) or used to clone individuals to bring the extinct species back into existence, or increase populations that are close to extinction.
3. Possible causes: single Chicxulub meteor impact in Yucatán, Mexico; multiple meteor impacts; massive volcanism in Deccan Traps, India; climate changes including rapid temperature increase followed by ice age. Answers should include analysis of the arguments.

1.14 Checking students' progress

The 'Checking your progress' section in the Student Book indicates the key ideas developed in this chapter and shows how students progress to more complex levels. It is provided to support students in:

- identifying those ideas
- developing a sense of their current level of understanding
- developing a sense of what the next steps in their learning are.

It is designed either to be used at the end of a chapter to support an overall view of the progress, or alternatively during the teaching of the unit. The students can self assess or peer assess using this as a basis.

It would be helpful if students can be encouraged to provide evidence from their understanding or their notes to support their judgements. In some cases it may be useful to explore the difference in the descriptors for a particular idea so that students can see what makes for a 'higher outcome'.

It may be useful with some descriptors to provide examples from the specific work done, such as an experiment undertaken or an explanation developed and recorded. If marking and feedback uses similar ideas and phrases, this will enable students to relate specific marking to a more general sense of progress.

To make good progress in understanding science students need to focus on these ideas and skills:

Students who are making modest progress will be able to:	**Students who are making good progress will be able to:**	**Students who are making excellent progress will be able to:**
Give some examples of differences between similar species and explain how these are used to classify organisms.	Explain the importance of the diversity of living organisms to life on Earth and why we have a common system for naming organisms.	Explain how scientists can use the universal system of classification to research or discuss an organism and to understand ecological relationships between organisms.
Identify some features of different organisms that are inherited and some that are determined by the environment in which the organism lives.	Explain the difference between continuous and discontinuous variation; explain why offspring from the same parents can be very different.	Use ideas and evidence to evaluate the importance of genetic and environmental variation.
Describe what selective breeding is and explain that it has produced new breeds of an organism.	Explain the process of selective breeding and why new breeds have been produced.	Explore and evaluate the advantages and disadvantages of selective breeding.
Identify examples of how variation causes competition for resources and causes natural selection.	Explore the theories of Lamarck, Wallace and Darwin, and explain their theories about why some organisms are better able to survive than others.	Evaluate the impact of Darwin's theories on contemporary science.
Describe chromosomes and their role in transferring heredity information to offspring.	Explain the relationship between chromosomes, genes and DNA; explain why offspring of the same parents may look very different.	Explore the role of scientists in the discovery of DNA and evaluate the relative importance of their contributions.
Describe how fertilised egg cells contain half of the chromosomes from each parent with a random mix of genetic information from each parent.	Explain how every new individual produced by sexual reproduction is genetically unique.	Explain the impact of slight 'changes' to DNA passed on from parents to offspring.
Describe cloning as one parent producing new individuals and identify examples of cloning that occur naturally; describe natural cloning as asexual reproduction.	Explain how artificial cloning is performed – for example in the creation of Dolly the sheep.	Explore and evaluate the advantages and disadvantages of artificial cloning; compare and contrast asexual and sexual reproduction.
Identify natural and human-caused environmental changes that have caused some species to become extinct.	Explain how the use of gene banks to preserve heredity material may prevent some endangered species from becoming extinct.	Analyse and evaluate the available evidence to explain why the dinosaurs suffered mass extinction.

1.15 Answers to Student Book questions

This table provides answers to the Questions section at the end of Chapter 1 of the Student Book. It also shows how different questions assess attainment in terms of the focus and style of a question as well as the context. Question level analysis can indicate students' proficiency in approaching different aspects of scientific understanding and different types of answer.

Q	Answer		Focus			Style			Context			
		Marks available	Knowledge & understanding	Application	Evaluation of evidence	Objective test question	Short written answer	Longer written answer	Variation & natural selection	Selective breeding	Chromosomes, genes & DNA	How features are inherited
1	a	1	x			x				x		
2	b	1	x			x					x	
3	d	1	x			x			x			
4	c	1	x			x			x			
5	• milk production	1	x				x			x		
	• meat production	1	x				x			x		
6	• stores	1	x				x				x	
	• containing frozen genetic material (from endangered species)	1	x				x				x	
7	• a species with no variation is vulnerable if there is an adverse change – e.g. new disease, change in the environment	1	x					x	x			
	• if all organisms are not adapted, the species will die out	1	x					x	x			
	• variation means some organisms have a favourable feature	1	x					x	x			
	• this gives them a survival advantage	1	x					x	x			
8	d	1		x		x			x			
9	b	1		x		x					x	
10	b	1		x		x					x	
11	any disadvantageous feature for a predator – e.g. • albinism • blunt teeth • poor sense of hearing and smell etc.	1		x		x			x			

Q	Answer		Focus			Style			Context			
		Marks available	Knowledge & understanding	Application	Evaluation of evidence	Objective test question	Short written answer	Longer written answer	Variation & natural selection	Selective breeding	Chromosomes, genes & DNA	How features are inherited
12	• so that when the egg and sperm cells fuse	1		x			x				x	
	• the fertilised egg has the normal number of chromosomes	1		x			x				x	
13	• extract the DNA from Megan's body cells	1		x			x					x
	• analyse to see if Megan has the gene for the disorder	1		x			x					x
14	• take a rose with large scented flowers	1		x				x		x		
	• cross it with a thornless rose	1		x				x		x		
	• select the plants with the fewest thorns/no thorns and the largest scented flowers to breed these together	1		x				x		x		
	• repeat for generations until he has a thornless rose with scented blooms	1		x				x		x		
15	• they have one gene for brown coat and one gene for white coat	1			x		x					x
	• brown colour is dominant over white so most of the offspring are brown, but a few are white	1			x		x					x
16	a) 118	1			x			x	x			
	b) size 6 to 16	1			x			x	x			
	c) continuous	1			x			x	x			
	d) 11	1			x			x	x			
	Total possible:	**30**	**12**	**12**	**6**	**8**	**10**	**12**	**12**	**7**	**7**	**4**

2.1 Introduction

When and how to use these pages

The Introduction in the Student Book indicates some of the ideas and skills in this topic area that students will already have met from KS2 or from previous KS3 work, and provides an indication of what they will be studying in this chapter. *Ideas you have met before* is not intended to comprehensively summarise of all the prior ideas, but rather to point out a few of the key ones and to support the view that scientific understanding is progressive. Even though students might be meeting contexts that are new to them, they can often use existing ideas to start to make sense of them.

In this chapter you will find out indicates some of the new ideas that the chapter will introduce. Again, it isn't a detailed summary of content or even an index page. Its purpose is more to act as a 'trailer' and generate some interest.

The outcomes, then, will be recognition of prior learning that can be built on, and interest in finding out more.

There are a number of ways this can be used. You might, for example:

- Use *Ideas you have met before* as the basis for a revision lesson as you start the first new topic.
- Use *Ideas you have met before* as the centre of spider diagrams, to which students can add examples, experiments they might have done previously or what they found interesting.
- Make a note of any unfamiliar/difficult terms and return to these in the relevant lessons.
- Use *In this chapter you will find out* to ask students questions such as:
  - Why is this important?
  - How could it be used?
  - What might we be doing in this topic?

Overview of the chapter

In this chapter, students will learn about drugs and diseases, and ways that the body is affected by each. The students consider the physiological effects of drugs on the body, in both the short and long term. They consider the wider implications of drug-taking on the individual and on society. Both legal and illegal drugs are studied.

The students learn about the causes of disease, including some historical events in both the cause and treatment of disease. They learn about how diseases are spread and then deduce how the spread could be reduced. They learn about the immune responses in the body and how scientists have learned to develop antibiotics and vaccines from these natural responses. The students are encouraged to consider the advantages and disadvantages of such medical treatments for individuals and society.

This chapter offers the opportunity for students to investigate bacteria and the effect of antibiotics. This is done by growing bacteria in the laboratory, giving the students the opportunity to understand and practise sterile technique. This is a good opportunity for students to understand how microbial scientists work.

Obstacles to learning

Students may need extra guidance with the following misconceptions:

- **Effects of drugs** All drugs are illegal. All drugs are bad for us. You will not be charged by the police for possessing drugs for your own use. Overcoming addiction requires only willpower. If recovery from addiction has failed in the past, it will never work.
- **Cigarettes and alcohol** Everyone who smokes will get cancer. Non-smokers cannot be harmed by smoking. Smoking cannabis is not as harmful as smoking cigarettes. Cannabis isn't addictive. Only alcoholics damage their bodies with alcohol.
- **Disease** All diseases are infectious. All diseases show symptoms. All microbes cause disease. As bacteria grow, they get bigger and bigger, and this is why bacteria can be seen on an agar plate.

- **Preventing and treating infection** Antibiotics and antibodies are the same thing. Disinfectants and antiseptics are the same thing. A vaccine is infection with the disease-causing microbe. Antibiotics can kill viruses. Vaccines cause autism.

	Topic title:	**Overarching objectives:**
2	Exploring types of drugs	The effects of 'recreational' drugs (including substance misuse) on behaviour, health and life processes
3	Understanding the impact of smoking	The effects of 'recreational' drugs (including substance misuse) on behaviour, health and life processes The impact of exercise, asthma and smoking on the human gas-exchange system
4	Considering the dangers of cannabis	
5	Understanding the effects of alcohol	The effects of 'recreational' drugs (including substance misuse) on behaviour, health and life processes
6	Exploring the effects of other drugs	
7	Learning about addiction	
9	Understanding how diseases are spread	*This part of this chapter is intended to embed and develop ideas from earlier in the Key Stage 3 course, including cell structure and function, micro-organisms and body systems.*
10	Exploring the body's defences	
11	Comparing microbes	
12	Investigating the growth of bacteria	
13	Understanding how antibiotics work	
14	Learning about vaccination	

2.2 Exploring types of drugs

Lesson overview

Learning objectives

- State examples of the four main groups of drugs.
- Describe the effects of different types of drugs on the body.
- Explain the effects of each type of drug on the body.

Learning outcomes

- Describe some types of drugs, with examples. [O1]
- Describe the effects of different types of drugs on the body. [O2]
- Explain the effects of different drugs on the body, including harmful effects. [O3]

Skills development

- Thinking scientifically: ask questions
- Working scientifically: n/a
- Learner development: act responsibly

Resources needed large sheets of paper; information about stimulants, depressants, painkillers and hallucinogens in sections (for jigsaw activity); Worksheet 3.2.2

Digital resources Quick starter; Slideshow: Exploring types of drugs - Information about the four main types of drug; Interactive activity: Drag the statements into the correct group - stimulant, depressant, painkiller or hallucinogen; Video

Common misconceptions All drugs are illegal. All effects of drugs are bad for us.

Key vocabulary painkiller, stimulant, depressant, hallucinogen, prescription drug, recreational drug, side effect

Teaching and learning

Engage

- **Group work** Ask the students to work in small groups to **compile** a list of drugs. They should work on large sheets of paper. [O1]
- Ask one group to **share** their ideas by displaying their work. [O1]
- Then ask other groups if they have any other drugs to suggest. Add these suggestions to the first display to compile a class list. [O1]

Challenge and develop

- Explain to the students that there are four main types of drugs – painkillers, stimulants, depressants and hallucinogens. [O1]
- Allow the students two minutes to **think about** what the characteristics of each group may be. Take feedback across the class, elucidating that painkillers stop or reduce pain, that stimulants speed up the body, that depressants slow down the body and that hallucinogens cause people to see or feel things that do not exist. [O1]
- Using task 1 of Worksheet 3.2.2 as a stimulus, the students work independently to **assign** examples of drugs to the appropriate groups. [O1]
- Ask the students to add to their lists other examples of drugs that were shared in the starter activity. [O1]

Explain

- **Jigsaw activity** Arrange the students in groups of four for a jigsaw activity. Give each student a number 1–4. Number 1 students **study** painkillers, number 2 study depressants, number 3 study stimulants and number 4 study hallucinogens, [O1, 2&3]

  Differentiation can be applied by assigning lower-attaining students number 1 or 2.

- The original groups should then split into 'expert groups' made up of students assigned the same number across the class. Provide each group with relevant information that uses both text and illustrations. Allow 10 minutes to **assimilate** information and **write notes**, of up to only 20 words and supported by diagrams, using task 2 of Worksheet 3.2.2. [O1, 2&3]
- The original groups re-form and each student takes two minutes to **teach** the group what they have learned. The rest of the group should complete task 3 of Worksheet 3.2.2 as a summary of their work. [O1, 2&3]

Consolidate and apply

- The students work individually to **produce a leaflet** to inform people about the effects of drugs. Encourage students to focus on the information included in the leaflet, rather than the layout or artwork – if necessary, you could provide a template that students could add text to. Remind them about the learning outcomes of the lesson and encourage them to try to demonstrate each of these in the leaflet. [O1, 2&3]

  You can differentiate this task by ensuring that higher-attaining students focus on learning outcome 3 as priority, giving less time to learning outcomes 1 and 2.

- **Pair work** Pair up students of similar ability. Ask them to **peer-assess** each other's leaflet, commenting on whether or not each of the learning outcomes has been met. Each student should then **identify** two positive points (2 stars) and an area for improvement (one wish). [O1, 2&3]

Extend

Ask students able to progress further to **consider**:

- the examples of drugs featured in the lesson – they should try to identify drugs that fall into more than one category
- what consequence this has on how the drugs may affect the body. [O1, 2&3]

Plenary suggestions

Learning triangle The students work individually to construct a learning triangle. They should draw a large triangle with a smaller inverted triangle that just fits inside (making four triangles). In the outer triangles, the students write something they've seen, something they've done and something they've discussed. Then, in the central triangle, they write something they have learned. [O1, 2&3]

Answers to Student Book questions

1. *painkillers*, e.g. paracetamol, codeine, morphine; *stimulants*, e.g. caffeine, ecstasy, amphetamines; *depressants*, e.g. alcohol, tranquillisers (sleeping tablets), heroin; *hallucinogens*, e.g. LSD, psilocybin mushrooms
2. *Prescription drugs* are prescribed by a doctor for medical conditions. *Recreational drugs* are taken for the feeling they give. 3. they may like the feeling; they may have problems that are trying to switch off from; they may be addicted etc.
4. caffeine is a stimulant; it 'increases' energy 5. it is a depressant; because it slows the body down
6. they are unwanted effects caused by drugs; e.g. antibiotics may cause diarrhoea (or other suitable examples)
7. a) there is a risk of heart attack because it increases the heart rate (or other risks such as dehydration because the taker becomes so energetic) b) there is a risk of depression; through anxiety because the drug affects chemicals in the brain

Answers to Worksheet 3.2.2

1. a) painkillers – paracetamol; stimulants – ecstasy, cocaine and, caffeine; depressants – alcohol, cannabis and morphine; hallucinogens – LSD and magic mushrooms b) other drugs added under the same (correct headings
2. summary of information about one group of drugs using symbols and/ or diagrams; 20 words maximum
3. summary of each drug group including the effect (e.g. stimulants increase heart rate), the side effects (e.g. stimulants can cause heart attack), their use (if any) (e.g. some depressants are used as anaesthetics)

2.3 Understanding the impact of smoking

Lesson overview

Learning objectives

- Describe the effects of smoking on the body.
- Explain the risks of smoking on the body.
- Examine the link between smoking and cancer.

Learning outcomes

- Identify parts of the body damaged by smoking. [O1]
- Explain how the body is damaged by smoking, including passive smoking. [O2]
- Analyse data about smoking and cancer to draw a conclusion and evaluate its reliability. [O3]

Skills development

- Thinking scientifically: analyse data
- Working scientifically: n/a
- Learner development: communicate effectively

Resources needed large sheets of paper and pens; body-torso model; Worksheet 3.2.3

Digital resources Quick starter; Interactive activity: Match the component of tobacco to its harmful effect on the body

Common misconceptions Everyone who smokes will get cancer. Non-smokers cannot be harmed by smoking.

Key vocabulary nicotine, tar, passive smoking, cancer

Teaching and learning

Engage

- **Group work** Ask the students to work in small groups to **discuss** how smoking affects the breathing system. This is an opportunity for students to revisit ideas from earlier work – such as tar builds up in alveoli and decreases gas exchange; tar causes cancer. The groups should **list** their ideas. [O1&2]
- Ask the groups to each take a turn in **sharing** one of their ideas. Keep going around the groups until there are no new ideas. [O1&2]
- Challenge the students to add the words 'nicotine',' tar' and 'carbon monoxide' to their lists if not already included – the links should be to the correct descriptions. Take some feedback across the class about the effects of each of these substances. [O1&2]

Challenge and develop

- **Group work** Ask the students to **draw** an outline of a human body on a large sheet of paper. They should then **highlight** the parts of the body that smoking can affect. [O1]
- Use the body-torso model and ask each group in turn to **describe** one part of the body that is affected by smoking. The students could come out and use the body model to demonstrate their point. [O1]
- Ask other groups to **explain** why each part of the body highlighted is affected by smoking. Encourage discussion between the presenting group and the groups chosen to explain the effect. [O1&2]

Explain

- As a class, ask the students to read the first and second sections in the Student Book. [O1&2]

- The students attempt tasks 1 and 2 of Worksheet 3.2.3. [O1&2]

  Ask higher-attaining students to add other examples to the list of descriptions and explanations.

Consolidate and apply

- **Pair work** Ask the students to read the 'Link with cancer' section in the Student Book on their own. Arrange the students in pairs and ask one student to **summarise** what they have read (without looking at the text). Their partner adds to the summary as necessary and then the pairs check the text again. [O3]
- The students attempt task 3 of Worksheet 3.2.3. [O3]

Extend

- Ask students who are able to progress further to **find out** about some of the support that is available to help people to stop smoking. They should **consider** how each type of support (e.g. nicotine patches, help groups and counselling, electronic cigarettes) can be helpful, as well as considering any disadvantages or dangers.

Plenary suggestions

Asking questions The students work in pairs to **devise a question** linked with the data on Worksheet 3.2.3. Encourage them to design questions worth 2 marks or more. They should also create a mark scheme before testing questions out on pairs. [O3]

Answers to Student Book questions

1. **three** from: heart increases rate/ more prone to heart attack; reduced circulation, leading to skin receiving less oxygen so more wrinkles; lungs blocked with tar more prone to infections; cilia paralysed by smoke so dust enters lungs
2. *nicotine* – addictive/ increases heart rate; *tar* – clogs up alveoli, leading to breathlessness/ more infections
3. smoke breathed in from another person's cigarette
4. by not smoking near non-smokers; not smoking in confined spaces (e.g. cars, houses); opening a window would reduce the effects but not all the smoke would be removed
5. smoke remains in the air for about two hours; particles spread and move from one room to others
6. nicotine is addictive; making it difficult to give up; some people may believe that cancer won't affect them
7. there are genetic factors; some people's genes mean that they are not as prone to cancer as others
8. 32 000 cases

Answers to Worksheet 3.2.3

1. nicotine; heart; overworked; attack; tar; gas; breathless

2 a) Nicotine increases the heart rate and the heart can be overworked; leading to a heart attack.

b) Cigarettes contain tar; tar is carcinogenic (as are other chemicals in cigarettes).

c) Cigarettes contain nicotine; this is an addictive drug.

3 a) must use data from the graph to support answers

i) Male and female percentages both decreased over time. Males started from higher percentage (approximately 65%) than females (approximately 40%) in 1950; but by 2010 approximately 20% of both males and females smoked.

ii) it decreased from approximately 110 per 100 000 in 1970; to approximately 60 per 100 000 in 2010 (in line with the decrease in percentage of smokers)

b) the line drawn should show a decrease in rate of lung cancer; a similar steepness to percentage of female smokers between 1970 and 2010

c) tobacco companies may be biased; because it is not good for business to highlight a link between cigarettes and cancer

2.4 Considering the dangers of cannabis

Lesson overview

Learning objectives

- Describe the medicinal uses for cannabis.
- Describe the negative effects of cannabis on the body.
- Give a balanced argument about whether cannabis should be legalised.

Learning outcomes

- Describe some reasons why people use cannabis. [O1]
- Describe the effects of cannabis on the body. [O2]
- Present an argument for or against the legalisation of cannabis, using evidence. [O3]

Skills development

- Thinking scientifically: consider the quality of evidence
- Working scientifically: interpret evidence
- Learner development: act responsibly

Resources needed mini-whiteboards (optional); additional information about legalising cannabis (optional); washing line or piece of string to stretch across the classroom; pegs (one per student); paper in two different colours; sticky notes; Worksheet 3.2.4 (the second page copied onto card)

Digital resources Quick starter; Interactive activity: Drag the statements into the correct group - for or against the legalisation of cannabis

Common misconceptions Smoking cannabis is not as harmful as smoking cigarettes. Cannabis is not addictive. You will not be charged by the police if you have cannabis for your own use.

Key vocabulary cannabis, illegal, legalise

Teaching and learning

Engage

- Ask the students to **decide** whether each statement in task 1 of Worksheet 3.2.4 is true or false. You could use mini-whiteboards for this activity or ask the students to work individually. [O1&2]
- Discuss each statement as a class and address any misconceptions. [O1&2]

Challenge and develop

- Give the students statements about the effects of cannabis. You could use those on task 2 in Worksheet 3.2.4. The students work in pairs to **sort** the statements in any way they choose – for example positive and negative effects, medical and recreational uses. [O1&2]
- **Pairs to fours** Each pair forms a four to **compare** how they sorted the effects. Encourage them to amend their sorting at this point if necessary. [O1&2]
- The students work individually to **summarise** each set they have created. [O1&2]

Explain

- Set up a continuum using a long piece of rope or string across the classroom. Add a sign saying 'Legalise cannabis' at one end and a sign saying 'Do not legalise cannabis' at the other. Ask the students to think for two minutes about whether or not cannabis should be legalised. Then ask them to hold a piece of paper with their name, positioning themselves along the continuum to **show their opinion** and to indicate how strongly they feel. Ask each student to peg their name where they are standing. You will return to this continuum later in the lesson. [O3]

- As a class, ask the students to read the second and third sections in the Student Book. You may also like to provide students with additional information about legalising cannabis – this could be found by searching 'should cannabis be legalised?' on the internet; information both for and against should be included. [O3]
- **Pair work** Ask the students to work in pairs to **organise the information** into statements for legalising cannabis and those against it – they could also refer back to information from the previous activity. Sticky notes could be used to write ideas on during this sorting activity. [O3]
- Ask groups to **share** the statements that are causing most disagreement and lead a discussion as to how these have been sorted by different groups. [O3]

Consolidate and apply

- The students work individually to **produce a persuasive piece of writing** for or against legalising cannabis. Task 3 of Worksheet 3.2.4 could be used as support for this task. [O3]

Extend

- Ask students able to progress further to identify people who may feel strongly one way or the other – for example an MS sufferer or the parent of a cannabis user. Ask them to **write a letter** from the viewpoint of their chosen role. [O1&3]

Plenary suggestions

Revisiting a continuum Return to the continuum and ask the students to reposition themselves in the appropriate place. They could add their names written on different coloured paper to show how their opinion has changed. [O3]

Changing opinions Ask any students who have moved to **share** a statement that changed their opinion (or the strength of their opinion). Ask students who have not moved to **respond** to these statements and **explain** why they did not change their opinion. [O3]

Answers to Student Book questions

1. depressant; hallucinogen
2. in multiple sclerosis, cancer, AIDS (or other sensible answer)
3. in multiple sclerosis; to decrease muscle spasm
4. tar
5. it causes anxiety, paranoia and schizophrenia; because it affects brain function
6. the comparison should include that: both contain tar (cannabis more); both contain other carcinogens (cannabis more); cannabis does not contain nicotine but tobacco does
7. Supplying affects more people than possession for personal use; the supplier is making money from an illegal activity.
8. dealers may compete; and threats and violence can occur between them; people can get into debt when dealers charge high prices; dealers may mix cannabis with other substances to make the drugs go further; and make more money and this can be dangerous
9. argument either for or against; with reasons

Answers to Worksheet 3.2.4

1. a) true b) false c) true (but only certain hospital doctors) d) false e) false f) true
2. a) There is no one correct way to sort the statements; students should be able to justify their choices; e.g. a), c) and h) together because they are medical uses of cannabis; e), i) and j) together because they are linked with how cannabis affects the brain.

   b) any sensible headings linked to their own chosen groups

   c) a summary of each of their chosen groups
3. a piece of writing that gives an opinion one way or the other; and includes information related to each of the bullet points in the checklist

2.5 Understanding the effects of alcohol

Lesson overview

Learning objectives

- Describe the short-term effects of alcohol on the body.
- Explain the long-term effects of alcohol.
- Suggest how alcoholism affects society.

Learning outcomes

- Describe some effects of alcohol on the body. [O1]
- Describe and explain several effects of alcohol on the body. [O2]
- Suggest and explain how alcoholism affects society. [O3]

Skills development

- Thinking scientifically: analyse data
- Working scientifically: present evidence
- Learner development: communicate effectively

Resources needed web-based clip of drunk person (optional); sticky notes; large sheets of paper; wine glass; shot glass; pint glass; empty spirits bottle; wine bottle; beer bottle; measuring cylinder; calculators; strips of coloured card or paper; Worksheet 3.2.5

Digital resources Quick starter; Interactive activity: Reorder the sentences to describe what happens to the body if a person continues to drink alcohol; Slideshow: Understanding the effects of alcohol - Information about alcohol and the dangers of long-term use; Video

Common misconceptions Only alcoholics do damage to their bodies with alcohol.

Key vocabulary alcohol, unit of alcohol, alcoholism, alcoholic

Teaching and learning

Engage

- Show a video clip of a drunk person – for example in a film or TV programme. Ask the students to **describe** what it is that suggests that the person has drunk alcohol. Alternatively, ask them to **discuss** in pairs how they would be able to tell that someone has drunk alcohol. Ask the students to **share** their ideas. Compile a list as a class. [O1]
- Ask the students to **produce a timeline** of the effects of alcohol on a person during a night's drinking. Ideas from Worksheet 3.2.5 could support this task. Each effect could be written on a sticky note before ordering. Ask the students to **compare** their timeline with that of another and then take feedback on any effects that were positioned differently. [O1&2]

Challenge and develop

- Ask the students to read the 'Effects of alcohol' section in the Student Book. [O1&2]
- **Pair work** The students should work in pairs to **recall** as much as possible. Ask them to cover up the page in the Student Book and recall what they have read. They should try to **reproduce** an annotated outline of the body – this could be done on large sheets of paper. They then uncover the page and check if they had forgotten any of the details. [O1&2]

Explain

- Using a measuring cylinder, measure out water to represent:
  - one small measure of vodka (25 cm^3) and pour it into a shot glass
  - a third of a pint of beer (190 cm^3) and pour it into a pint glass
  - half a glass of wine (87 cm^3) and pour it into a wine glass. [O2]

- Ask the students to **suggest** what these volumes of alcoholic drink have in common. Explain that each contains one unit of alcohol. Ask the students to **suggest** why the volumes are different for the same number of units. [O2]
- Show them a spirits bottle, a wine bottle and a beer bottle and ask them to **find** the alcohol percentage of each drink on the bottle. If necessary, discuss what is meant by percentage. Ask the students to **arrange** the bottles in order of decreasing alcohol content. [O2]
- Ask them to read the 'Long-term damage' section in the Student Book. Use questions and answers to ensure understanding of the unit system for alcoholic drinks. [O2]
- The students work individually to attempt task 2 of Worksheet 3.2.5. [O2]

Consolidate and apply

- **Pairs/spies** Ask the students to work in pairs to **identify** how alcoholism impacts on society. The pairs could use the 'How does alcohol affect society' section in the Student Book as a stimulus for this activity. They should write each idea on a separate piece of coloured card or paper. One student should then act as a spy and visit two other groups. They then return to their original partner and **share** any new ideas. [O3]
- Provide the students with headings to write about, for example, the effects on police, families, crime and the NHS. They should then **arrange** their ideas in a logical order to begin to **structure a piece of writing** to describe and explain the impact of alcohol on society. Encourage the students to use a paragraph for each idea. Encourage them to include an opening and closing paragraph or statement. [O3]

  Allow lower-attaining students to verbalise their ideas or to use bullet points, rather than writing continuous prose.

Extend

- Ask students able to progress further to **consider** why alcohol affects different people in different ways – for example two people consuming the same amount of alcohol. The students should **compile** a list of possible factors and write an **explanation** to support each idea. [O1&2]

Plenary suggestions

Facts and opinions Ask each student to **share one fact** and **one opinion** based on this lesson. Ask other students to comment if a point is not identified correctly as fact or opinion.

Answers to Student Book questions

1. feeling more confident; it slows the brain so may be unbalanced; speech may be slurred; people react less quickly (or other sensible answers)
2. Alcohol suppresses the part of the brain that inhibits us.
3. the heart rate and breathing rate decrease; the nervous system then fails
4. 9
5. a) 9 b) $2\frac{1}{3}$ (2 bottles and 1 large glass)
6. the table should include the effects on the brain (depression); heart (increased heart disease and blood pressure); stomach (increased risk of stomach cancer); liver (liver disease and jaundice)
7. the children or partner may suffer violence or neglect; there may be financial issues through needing money for alcohol or because of not working; the effects of seeing a family member suffering the effects of alcohol
8. sensible answers including reference to the effect on families, workplaces, crime, health and healthcare

Answers to Worksheet 3.2.5

1. some answers could be on the same place in the timeline – suggestions are shown in brackets or these effects may be swapped in order: (more confident, talkative); (slur speech, lose balance, fall over); vomit; coma; death
2. a) 15 units b) i) 1.4 units ii) 9.75 units

   c) so that we can stay within recommended limits; to avoid too much damage to the body by alcohol
3. information can be found on the internet; the students should include a reference to the importance of helping people to overcome alcoholism and to reduce the damage caused to society (e.g. to families, the NHS)

2.6 Exploring the effects of other drugs

Lesson overview

Learning objectives

- Describe the effects of different drugs on the body.
- Compare the dangers of different drugs.

Learning outcomes

- Describe the effects of drugs such as ecstasy, heroin and cocaine on the body. [O1]
- Describe both the short-term and long-term effects of ecstasy, heroin and cocaine. [O2]
- Compare the effects of different drugs on the body. [O3]

Skills development

- Thinking scientifically: ask questions
- Working scientifically: n/a
- Learner development: ask questions

Resources needed images linked with ecstasy, cocaine and heroin (e.g. ravers, dancing in a club, cocaine being snorted, drugs being injected, white powder, heroin being heated on a spoon, tablets supposed to be ecstasy); separate information about ecstasy, cocaine and heroin (for example from the Student Book, other textbooks and the internet for workstations); stopwatch or stopclock; coloured strips of paper; Worksheet 3.2.6

Digital resources Quickstarter; Interactive activity: Drag the drugs into the correct group - legal or illegal

Key vocabulary class A drug, ecstasy, cocaine, heroin

Teaching and learning

Engage

- Ask the students to look at some images linked with ecstasy, cocaine and heroin (although do not tell them the link at this stage). These could include ravers, dancing in a club, cocaine being snorted, drugs being injected, white powder, heroin being heated on a spoon and tablets supposed to be ecstasy. Ask them to **suggest** what the lesson is about. [O1]
- **Group work** Once the focus of the lesson has been identified as being ecstasy, cocaine and heroin, ask the students to **discuss** in small groups why these three drugs have been specified together. Take some **suggestions** from groups and ensure that they all realise that these are the three most widely used illegal drugs after cannabis. [O1]

Challenge and develop

- Ask the groups to **identify** the type of drug that each of ecstasy, cocaine and heroin is. They should then **list** the effects of those groups of drugs, including the harmful effects. The groups then **discuss** their answers with another group and add to their own work if necessary. [O1&2]
- As a class, read the 'Effects of ecstasy, cocaine and heroin' section in the Student Book. Check the understanding of any key words and scientific terminology as you read. [O1&2]
- Ask the students to read the 'Size of the problem' section in the Student Book individually and answer questions 4 and 5. [O1&2]

  This activity could be differentiated by asking higher-attaining students to **plot a graph** of the data in Table 3.2.6 before answering question 5.

Explain

- Around the classroom, set up work stations with the information suggested under *Resources needed* – at least one for each of the drugs ecstasy, cocaine and heroin. The students visit each station and **assimilate** the information. They should be allowed to spend five minutes at each work station. [O1&2]
- Allow the students time to use the information they gathered to **construct a table** to show the short-term effects and long-term effects of each of the three drugs. [O1&2]

Consolidate and apply

- **Pair talk** Working in pairs, the students **devise questions** about the three drugs, along with a mark scheme. Encourage them to use 'compare' as the command word in one of their questions. Worksheet 3.2.6 could be used here to provide the students with background and examples of questions. [O1, 2&3]
- **Pairs to fours** The pairs form fours to test their questions, with each pair **giving feedback** in the form of two stars and a wish. [O1, 2&3]

Extend

- Ask students able to progress further to **give an opinion** as to the most and least dangerous of the three drugs discussed in this lesson. As many points as possible should be given to **justify their opinion**, with evidence presented wherever possible. [O1, 2&3]

Plenary suggestions

Loop card game Ask each student to **write a question** linked with the lesson on a strip of coloured paper. They then **write the answer** on a different- coloured strip of paper. Arrange the students in groups of 6–8 and hand out a question and answer (randomly) to each. One student reads out their question – the student holding the correct answer reads it out and then reads their question out and so on. [O1, 2&3]

Answers to Student Book questions

1. increased heart rate; body temperature; may feel depressed and flu-like when the effects wear off
2. the throat is relaxed; and so the person is not able to cough vomit up; they could choke
3. there is an increased heart rate, which could lead to heart attack
4. drug users may die from related illnesses; difficult to prove death caused by drug-taking; some deaths are not explained
5. Explanation to include: most deaths caused by heroin alone, then cocaine, then ecstasy. All drugs are linked with deaths involving other drugs as well. The student may offer some numerical analysis such as deaths caused by other drugs as well; 3 times as many for ecstasy; 7 times as many for cocaine; 2.5 times as many for heroin. They may make a comment about cocaine leading to other drug use, more so than ecstasy or heroin.
6. they have the same prison sentence; injecting both cocaine and heroin can cause ulcers and collapsing of veins; cocaine and heroin are both highly addictive; ecstasy and cocaine both increase the heart rate and cause heart problems; ecstasy and cocaine both cause depression
7. agreeing or disagreeing is acceptable; must be backed up by explanation, such as heroin causes more deaths and so should have longer prison sentence
8. any opinion is acceptable, so long as reasons are given; for example dangers concerning heart attacks and coma are most important because they can lead to death

Answers to Worksheet 3.2.6

1. a) i) stimulant ii) stimulant iii) depressant

   b) i) ecstasy and cocaine ii) heroin iii) cocaine and heroin iv) ecstasy and cocaine
2. a) Most harmful: heroin, cocaine, ecstasy and LSD; life imprisonment for selling; possessing heroin or cocaine could lead to a prison sentence of up to 7 years.

   b) Amphetamines and cannabis; possessing cannabis may lead to an on-the-spot fine or a warning.

   c) Tranquillisers and anabolic steroids; could receive 14 years in prison for selling.
3. a) ecstasy b) 16–19, 20–24 and 25–29; because the use of all three drugs is the highest

   c) Ecstasy use increases from age 16–19 to age 25–29; but then decreases. It is almost zero use at age 55–59. Credit any numerical data from graph. Cocaine use follows similar trend to ecstasy; although still some cocaine use at age 55–59. Credit any numerical data from graph.

   d) A higher proportion of 16–19 year olds use cannabis than 20–24 year olds. However, a higher proportion of 20–24 year olds use cocaine and ecstasy than 16–19 year olds. This may be because cannabis is more freely available in places where younger people go; or because cannabis is cheaper than cocaine and ecstasy, and the higher age group have more money.2.7 Learning about addiction

2.7 Learning about addiction

Lesson overview

Learning objectives

- Define addiction.
- Describe how drugs affect the brain.
- Explain the effects of withdrawal on the body.

Learning outcomes

- Describe addiction, with examples. [O1]
- Describe the changes in the brain caused by drugs and explain how this can lead to addiction. [O2]
- Explain the effects of withdrawal on the body, including the brain. [O3]

Skills development

- Thinking scientifically: ask questions
- Working scientifically: n/a
- Learner development: collaborate effectively

Resources needed images of different examples of addiction– e.g. alcohol, drugs such as heroin, cocaine, gambling; large sheets of paper and pens; sticky notes (different colour for each group); Worksheet 3.2.7 (the second page copied onto card)

Digital resources Quick starter; Interactive activity: Drag the withdrawal symptoms into the correct group - emotional or physical; Hangman: Key vocabulary game

Common misconceptions Overcoming addiction is just a case of having willpower. If recovery from addiction has failed in the past, it will never work.

Key vocabulary addiction, withdrawal, rehabilitation

Teaching and learning

Engage

- Show pictures (possibly as a slideshow) of things that people can be addicted to – for example, different drugs, alcohol, cigarettes, gambling. Ask the students to **suggest** what the lesson is about and then to **identify** each addiction. [O1]
- **Group work** Working in small groups, ask the students to **discuss** what 'addiction' is. Take feedback from one group and then ask other groups to amend or extend the explanation. Bounce ideas around the classroom until addiction is defined as 'someone who craves whatever they are addicted to mentally or physically'. [O1]
- The students could then attempt task 1 of Worksheet 3.2.7. [O1]

Challenge and develop

- **Group work** Arrange the students in groups of three. Ask the groups to **discuss** why people become addicts. You may want to use a specific example, such as a heroin addict or an alcoholic. The groups **compile a list** of factors that promote addiction. [O1]
- The groups join together and **share** their lists. They should **discuss** any different factors that they have listed. As groups work, make a note of common factors identified – share these with the whole class. [O1]
- Read the list in the 'What is addiction?' section in the Student Book as a class and discuss any points not included in the students' lists. [O1]

Explain

- **Pair talk** Ask the students to read the 'Changes in the brain' section in the Student Book. In pairs, the students take turns to **recall** and **explain** what the text is about. [O2]
- The students work individually on task 2 of Worksheet 3.2.7. [O2]

Consolidate and apply

- The students work in pairs to **draw a mind map** with the word 'addiction' at the centre. Encourage the students to include as many different ideas as possible, and to include only one word on each branch of the mind map. They could use the 'Overcoming addiction' section in the Student Book to support this task. [O1, 2&3]
- Ask the students to visit other mind maps. Each group should add a **comment** to each map on a sticky note (each group should be given a different colour). Ensure that the comment is formative – for example suggesting another link or keyword to include. [O1, 2&3]
- The students could attempt task 3 of Worksheet 3.2.7 at this point. [O3]

Extend

- Ask students able to progress further to **consider** how rehabilitation clinics should be funded. They should **give their opinion** on whether or not public money should fund support for drug addicts, **providing explanations** for their opinions. If possible, pair students up to **discuss** their thoughts. If they disagreed, the aim is to persuade their partner that their idea is valid.

Plenary suggestions

Freeze frame Ask the students to create a freeze frame of an idea from the lesson. This involves three or four students arranging themselves as static images. Other students suggest what it represents.

Answers to Student Book questions

1. when the body becomes dependent on a drug, and the person craves the drug/change in brain.
2. nicotine; alcohol; cannabis; sedatives; cocaine; heroin
3. it depends on factors such as: genetics; previous trauma; mental health issues; the age when drugs were first tried; and the way that drug is taken
4. they fool the brain because they are similar to brain chemicals; the brain then sends abnormal messages
5. they get a good feeling when they take cocaine, because dopamine is released over the brain; they take the drug again to try to repeat that feeling
6. willpower alone may not be enough for an addict to recover; physical changes in the brain take place, not just behavioural changes
7. withdrawal should be done gradually and in a controlled way; some of the withdrawal symptoms are dangerous and may need monitoring
8. Rehabilitation takes an addict away from their routine and friends, and so starts to break habits. Counselling allows an addict to change their thinking and to identify triggers. A doctor will be present to monitor and treat physical symptoms.

Answers to Worksheet 3.2.7

1. alcohol/cocaine (either way); normal; stealing; damage; craving; resist
2. a) *Statement A*: If the drug is not taken by an addict, withdrawal symptoms are experienced. In an addict, the drug is taken to feel normal. Without the drug, the body cannot behave normally.

   Statement B: many drug addicts risk going to prison to get money for drugs by stealing. Some drugs suppress the part of the brain that is needed to control behaviour and make decisions.

   Statement C: mental health problems such as paranoia and schizophrenia are more common in drug users than non-drug users. Some drugs interfere with the release of chemicals in the brain. This can cause long-term damage.

   b) *Explanation*: There are genes that make us susceptible to addiction and genes are passed on in families (from mother and father to child).

   Evidence: Many drug addicts have relatives such as parents, brother/sisters who were also users or addicts.
3. information to include examples of physical and emotional withdrawal symptoms; medical help needed because it could be dangerous to stop too quickly due to possible breathing difficulties; rehabilitation clinics can remove an addict from the social setting and so help to break the habit as well as give medical attention

2.8 Applying key ideas

Driving under the influence

Objectives

- Extract information about the effects of alcohol and other drugs on drivers.
- Consider reasons why it is difficult to gather evidence about the effects of drugs.
- Use ideas and information about driving under the influence to form opinions and make reasoned arguments about solutions and punishments.

Outcome

- Clear and effective responses to questions, indicating understanding and the next steps in learning.

The purpose of this activity is to provide an opportunity to see how successfully students are grasping the key ideas so far. It isn't designed to be used as a formal test – it might be that the students work on the questions collectively. It does provide an opportunity for the teacher to look at written work, engage students in discussions and form ideas about progress being made.

The tasks are progressive. Lower-attaining students should be able to tackle the first two or three tasks and higher-attaining students will find the later ones more challenging.

Resources needed pictures of road traffic accidents caused by drink-driving (or web-based drink-driving advert); optional web-based video clip to show field impairment test; Worksheet 3.2.8

Teaching and learning

Engage

- Challenge the students to **reflect back** over the last few lessons and **identify** some of the key ideas they've met. Draw out some responses about these ideas and say that it's important as learners that they develop a sense of their own progress.
- Show some pictures of road traffic accidents or a web-based drink-driving advert. Ask students to **suggest** some causes of road traffic accidents.

Challenge and develop

- Arrange the students in small groups. Ask them to **revisit** the effects of drugs such as alcohol, cannabis and ecstasy on the body – they may want to use the Student Book to help with this activity.
- Ask them to **compare** their lists with another group and to add to their own work or **seek clarification** if necessary. Next ask the groups to focus on which of these effects would be important if planning to drive.

Explain

Ask the students to read the 'Driving under the influence' in the Student Book. Depending on their ability to assimilate text, you may need to adopt strategies such as:

- Have the students working in pairs, with copies of the text (provided as Worksheet 3.2.8) and highlighters. One student highlights key ideas and the other (using a different colour) any words they don't understand.
- Using the text as a source of ideas, draw a spider chart with 'Driving under the influence' at the centre and the causes, effects and consequences around it.
- Work with any group of students that have difficulty with reading to support them in decoding the text and accessing the ideas.

Consolidate and apply

- Now ask the students to attempt the Student Book tasks – they might do this individually or collaboratively. In each case, encourage them to identify and record their ideas.
- Ask the students to **present their responses** to various tasks, either orally or by displaying their work.

Extend

- For some students the later tasks (particularly 5 and 6) give opportunities for extension work.

Plenary

Ask the students for responses to questions such as:

- How well were you able to approach those tasks?
- Did you manage to present some high-quality responses?
- Are there some ideas that you need to strengthen in the later part of this topic?

If there are, make a note.

Provide the class with some overall feedback, indicating ideas that they have grasped effectively and those that may need developing further.

Likely responses and next steps in learning

1. This task gives the students an opportunity to search text and to extract specific information. This task also requires the students to compare methods – students may comment that testing for alcohol is more scientific or easier to prove than for other drugs.
2. This task involves recalling that different drugs have many different effects, and that some of these effects are difficult to spot or prove. Students may recall from the 'challenge' activity that ecstasy causes an active state, whereas cannabis causes a more relaxed state, or that cannabis causes bloodshot eyes whereas ecstasy does not. The students may recognise that these effects are only signs, rather than definitive proof.
3. This task allows students to consider why drugs affect people in different ways. They should discuss factors such as the mass of the individual, the gender (alcohol is broken down less quickly by females), how much the person has eaten. Other ideas may also be given and each should be clarified. Students are asked to make a suggestion and so any sensible suggestion, with reasons, should be given credit. For example, they may point out that even the smallest level of alcohol in the body may slow reactions and so it is safer for the limit to be lower.
4. This task requires students to convert numerical data into percentages. They should calculate that 10 per cent of alcohol drinkers admitted to driving while under the influence, whereas 20 per cent of drug users admitted to driving under the influence. Some students may also include the statistic that 1 per cent of drug users admitted to driving under the influence almost every day. Students may suggest that it is difficult to gather this information because people may be reluctant to admit it, even more so with other illegal drugs.
5. This activity requires students to use the information given (and that studied in previous lessons) to form their own opinion. Any opinion should be backed up with reasons and evidence where possible. Students may suggest that sanctions are successful because deaths caused by drink-driving have decreased. However, they may suggest that because there are still *some* deaths caused by drink-driving, and that those caused by drug driving are thought to be increasing, then sanctions are not successful. Other deterrents may include longer driving bans, increased fines or more radical suggestions such as having to meet the families of those killed by drivers under the influence.
6. Students are asked to write an argument in this task and are given guidance on how to do this – an argument should include ideas from both sides and should end with a personal conclusion. Arguments *for* random testing include that it may decrease the number of road accidents caused by drivers under the influence of drugs, it may reduce the number of people willing to take the risk, it may raise awareness of the problems of driving while under the influence of drugs. Arguments *against* include that it could be expensive and difficult to manage, along with the problems of which drugs would be tested for and which we are able to definitively test for etc. Any conclusion is acceptable so long as it is justified in the preceding text.

2.9 Understanding how diseases are spread

Lesson overview

Learning objectives

- Describe how diseases are spread.
- Consider ways of reducing the spread of specific diseases.

Learning outcomes

- Describe some ways in which diseases are spread, with examples. [O1]
- Provide several examples of how specific diseases are spread and suggest how spread may be reduced. [O2]
- Consider suggestions to reduce the spread of specific diseases and justify plans. [O3]

Skills development

- Thinking scientifically: evaluate risks
- Working scientifically: develop explanations
- Learner development: communicate effectively

Resources needed sticky notes; mini-whiteboards; Worksheet 3.2.9 (second page copied onto card)

Digital resources Quick starter; Slideshow: Preventing the spread of disease - A look at how infectious diseases are spread and how this can be prevented; Interactive activity: Match the way a disease is spread to the mechanism by which it can be prevented; Video

Common misconceptions All diseases are infectious. All diseases show symptoms.

Key vocabulary infectious disease, non-infectious disease, symptom

Teaching and learning

Engage

- **Group work** Arrange the students in small groups – each group **brainstorms** 'diseases'. One person in the group **scribes** the ideas, one person ensures that each group member takes a turn to **share** ideas and one person is responsible for **feeding back** to the class. Ask one group for feedback and then ask each group in turn to add ideas until there are no more to share. [O1]
- As groups, the students try to **sort** the diseases into those that can be 'caught' and those that cannot. Sticky notes could be used for this activity. As a class, sort the diseases. Introduce the terms 'infectious' and 'non-infectious'. Reinforce this by asking the students to attempt task 1 of Worksheet 3.2.9. [O1]

Challenge and develop

- Using the cards made in task 2 of Worksheet 3.2.9, ask the students to work in pairs to **match** each disease with its method of spreading and details about the method. Alternatively they could copy out the information into appropriate places in a table. Students should read the 'How do diseases spread?' section in the Student's Book to help with this activity. [O1&2]

  You could differentiate this task by providing additional information for higher-attaining students. Ask them to add other examples to their table and to identify examples of diseases that have more than one method of transmission. [O1&2]

Explain

- **Pair talk** Ask pairs to **discuss** a method of trying to reduce the spread of each of the diseases included in the previous activity. Take some feedback. As a class, read the 'Preventing the spread of disease' section in the Student Book and ask the students to **comment on** whether the class and the book ideas are similar or not. [O2&3]

- Ask the students to work individually to **present** the information about the spread of diseases and how to prevent the spread, in any way they choose. You may want to suggest examples such as a comic strip, poster or rap. [O2&3]

Consolidate and apply

- The students work individually to **explain** how each method used to prevent the spread of different diseases works. You may want to discuss with them what a good explanation includes before they begin this activity – for example an explanation may answer questions such as 'How?' and 'Why?' or it may use the word 'because'. [O2&3]
- The students work individually to **produce an educational leaflet** to include:
  - information about how diseases are spread and who is at risk
  - a description of how spread can be reduced
  - an explanation of how spread is reduced.

Extend

- Ask students able to progress further to **consider the importance** of education about the spread of diseases. They may want to consider the importance in developing countries as a specific example. Task 3 of Worksheet 3.2.9 could be used here. [O3]

Plenary suggestions

Quick fire questions Ask quick questions summarising the lesson – for example you name a disease and the students write down the method of transmission. Ask the students to **display answers** using mini-whiteboards. Ask them to take a turn in **asking questions** of the rest of the class. [O1, 2&3]

Answers to Student Book questions

1. infectious disease can be spread, non-infectious diseases cannot; infectious – chickenpox, influenza; non-infectious – cancer, diabetes
2. we may get symptoms, such as fever, rash and pain
3. sewage spills into water; the water is not filtered and purified as it is in developed countries, so microbes are not removed
4. children may not wash their hands well after using the toilet, so traces of urine and faeces remain and may pass to the child's mouth by hand or via an object
5. a) alcohol gel reduces the number of microbes that could cause infection; it is more practical than washing hands with soap each time we enter a ward; it is important because ill people in hospital more susceptible to infection

   b) blood is screened for infectious diseases that are passed through blood such as HIV, hepatitis; so is not passed on during/after transfusion. It also acts as a check for the donor
6. Dogs are not in contact with people or animals during quarantine. If the quarantined dog is clear of symptoms of rabies, it is certain that they do not carry the infection. People object because think that their animals were not at risk of carrying rabies, and may think it is cruel to keep dogs in quarantine.

Answers to Worksheet 3.2.9

1. a) infectious b) non-infectious c) chickenpox d) diabetes
2. a) Method A: disease is spread by eating contaminated water or food

   Method B: disease is spread by animal or insect bite

   Method C: disease is spread by blood from one person coming into contact with another, such as during a blood transfusion or when sharing needles

   Method D: disease is spread by airborne droplets from the mouth and nose, for example, when sneezing

   Method E: disease is spread by tiny amounts of urine or faeces passing from person to person or via an object

   Method F: disease is spread by mixing body fluids during sexual contact

   b) i) food/water ii) faeces/urine iii) sex and blood iv) animals v) air
3. The letter should be written in continuous prose; outlining ideas such as education could reduce number of deaths or serious illnesses in developing countries. The student may use a specific disease as an example; e.g. typhoid and cholera spread by drinking contaminated water.

2.10 Exploring the body's defences

Lesson overview

Learning objectives

- Describe how the body resists infection.
- Explain the role of white blood cells in fighting infection.

Learning outcomes

- Describe the body's mechanisms to prevent infection. [O1]
- Explain the roles of white blood cells in fighting infection. [O2]
- Explain why we rarely catch the same disease twice, but we may catch flu over and over again. [O3]

Skills development

- Thinking scientifically: understand how theories develop
- Working scientifically: n/a
- Learner development: plan progress

Resources needed body-torso model; large sheets of paper; images of blood cells – red blood cells, white blood cells and platelets; pre-prepared slides of blood cells for Bioviewer(s) or microscope(s) (optional); modelling clay of different colours; camera(s) or video camera(s); Worksheet 3.2.10

Digital resources Quick starter; Interactive activity: Match the body barrier to the way it prevents microbe entry

Key vocabulary white blood cell, immune system, phagocyte, antibody, memory cell

Teaching and learning

Engage

- **Think, pair, share** Show the body-torso model. Ask the students to **think** for two minutes about how the body prevents microbes from entering – they then **share** their ideas with a partner. Then take feedback across the whole class. [O1]
- **Pair work** Ask the students to work in pairs to **draw** an outline of the body on a large sheet of paper. They then **annotate** the diagram to **summarise** the ways in which the body prevents infection from entering. Alternatively they could attempt task 1 of Worksheet 3.2.10. [O1]
- Pairs swap diagrams, study their partner's, and then add extra information to their own diagram. [O1]

Challenge and develop

- Ask the students to **discuss** why we have blood. Take feedback from the class and make sure they all **understand** that one of the roles of blood is defence from infection. [O2]
- The students look at images of blood cells (printed pictures, projected images or using Bioviewers or microscopes). Ask them if they can **identify** any of the different blood cells. Point out some of the features – such as red blood cells with no nucleus, platelets being fragments of cells, white blood cells have a nucleus and are not regular in shape. [O2]
- Ask the students to **recap** the role of red blood cells in carrying oxygen. Explain the role of platelets in clotting blood (and forming scabs). Ask them to read the 'Defence by white blood cells' section in the Student Book. They should then work with a partner to **explain** to each other what they have read. Use verbal questions and answers to assess understanding. [O2]

Explain

- If possible, show an animation of how antibodies and antitoxins work. [O2]

- Working in pairs of similar ability, the students **draw** an annotated poster about white blood cells and the role of each type. Agree success criteria as a class before they begin the activity. For example:
  - they must include the terms 'immune system', 'white blood cells', 'phagocytes', 'B-cells' and 'T-cells'
  - the posters must include text about each type of white cell, as well as an illustration. [O2]
- Alternatively they could attempt task 2 of Worksheet 3.2.10. [O2]
- The pairs swap posters and **peer-assess** the posters using the success criteria agreed as a class. Feedback could be provided in terms of two stars and a wish. They then revisit their own poster and act on feedback if that is appropriate. [O2]

Consolidate and apply

- As a class, read the 'Memory cells' section in the Student Book. The students should then add to their posters from the previous activity. [O3]
- **Pair work** Arrange the students in pairs and ask each pair to **make a model** about the role of memory cells using modelling clay. This activity is outlined in task 3 of Worksheet 3.2.10. If possible, allow them to take a series of photographs or to **record** their animation. Choose some pairs to **demonstrate** their models. [O3]

Extend

- Ask students able to progress further to include in their model how influenza can cause infection again and again, without recognition by the immune system. [O3]

Plenary suggestions

Glossary Ask the students to **make a glossary** of key words used in this and the previous lesson. This could be written as a list to add to during the rest of the topic. They could produce this as a display. [O1, 2&3]

Answers to Student Book questions

1. to prevent microbes entering the body through the cut
2. mucus traps microbes; an enzyme in mucus may kill them or they may be swallowed and then destroyed by stomach acid
3. they are the first way that the body prevents infection; (the second line is the immune system)
4. white blood cells help to fight infection; so when someone has an infection, the cells increase in number
5. the blood flow to the site of infection increases
6. the diagrams should be annotated and include role of phagocytes, T-cells and B-cells
7. memory cells recognise the microbe (virus) causing measles and quickly destroy the microbes before infection can cause symptoms
8. the microbe changes the proteins on its surface
9. B-cells produce antibodies; T-cells co-ordinate other white blood cells to destroy the microbes

Answers to Worksheet 3.2.10

1. a) Skin prevents microbes entering the body.
   b) Mucus traps microbes (and has enzymes that destroy them).
   c) Tears wash microbes out of the eyes (and contain an enzyme that destroys them).
   d) Stomach acid kills many microbes (low pH).
   e) A scab prevents microbes from entering the body.
2. a) i) Phagocytes ingest and destroy microbes. ii) B-cells produce antibodies that attach to microbes and destroy them.
   iii) T-cells produce antitoxins that neutralise the toxins produced by microbes, and co-ordinate the attack of white blood cells on microbes.
   b) Any suitable suggestions; such as 'ph' for 'food' because phagocytes ingest as though they are eating microbes; B-cells have 'B' for antiBodies; T-cells have T for antiToxins.
3. a) suitable model showing that after an infection, some B-cells and T-cells remain; if the same microbe (in the same format) enters again, the memory cells increase in number; and antibodies and antitoxins are produced before infection can take hold
   b) model to show that the surface of the microbe has changed and is not recognised by the memory cells

2.11 Comparing microbes

Lesson overview

Learning objectives

- Describe the characteristics of different types of microbe.
- Recall examples of diseases caused by bacteria, viruses and fungi.
- Evaluate a model of a microbe.

Learning outcomes

- Describe the characteristics of different microbes. [O1]
- List some diseases caused by microbes, using examples. [O2]
- Create and evaluate a model of one type of microbe. [O3]

Skills development

- Thinking scientifically: use units and nomenclature
- Working scientifically: evaluate data
- Learner development: plan progress

Resources needed pictures of bacteria, viruses and fungi; mini-whiteboards; equipment and materials as detailed in the Technician's notes; Worksheet 3.2.11; Practical sheet 3.2.11; Technician's notes 3.2.11

Digital resources Quick starter; Slideshow: Types of microbe - A look at the features of fungi, viruses and bacteria; Video; Interactive activity: Drag the statements into the correct group - bacteria, virus or fungi

Key vocabulary fungus, bacteria, virus, micrometre

Teaching and learning

Engage

- Show pictures of bacteria, viruses and fungi (at least two of each). Ask the students to **describe** the pictures. They form pairs and **compare** their descriptions – ask some pairs to **share** examples of descriptive language. [O1]
- **Pair work** Show the pictures again and ask pairs to **group** the pictures into three groups. Ask them to **explain** why they grouped certain pictures together. Introduce the terms 'bacteria', 'viruses' and 'fungi' and discuss some of the features of each as a class. [O1]
- Students attempt task 1 of Worksheet 3.2.11. [O1]

Challenge and develop

- Read the 'Which microbe is to blame?' section in the Student Book as a class. Ask the students to answer questions 4, 5 and 6 of the Student Book verbally in pairs. [O2]
- Using a stopwatch or timer, allow the students 30 seconds to **memorise** the information in Table 3.2.11a. Ask questions about what causes different diseases. They could display their answers using mini-whiteboards. Ask the students to **reflect on** how they memorised the information and **share** strategies such as visualising a story or using mnemonics. [O2]
- Students attempt task 2 of Worksheet 3.2.11. [O2]

Explain

- Provide the students with a choice of craft and 'junk' items. Ask them to work in pairs to **make a model** of one type of microbe, following Practical sheet 3.2.11. Encourage them to focus on the features of each microbe identified in the 'engage' activity of the lesson. [O3]

- **Pairs to fours** Pairs merge to form fours and each pair **identifies** what each object in the other model represents. [O3]

Consolidate and apply

- Display all the models around the room. The students **examine** the models to **identify** which type of microbe it is representing. Encourage them to identify some good features of the other models. Pairs then **evaluate** their own model. Task 3 of Worksheet 3.2.11 could be used to structure this activity. [O3]

Extend

- Ask students able to progress further to **find out** about the relative sizes of different types of microbes. Then ask them to **reflect on** whether or not the proportions are reflected in the models made by themselves and other groups. [O1]

Plenary suggestions

What we have learned Ask the students to **consider** individually what they have learned during the lesson. Encourage them to consider skills as well as content here. Then ask them to **discuss** in pairs how making a model has helped their understanding. If it has not helped, ask them to **reflect on** what type of activity could be used instead. Take one comment from each student. [O1, 2&3]

Answers to Student Book questions

1. suitable table to show *fungi* have a cell wall, cell membrane and nucleus; *bacteria* have a cell wall, cell membrane, but no nucleus (circular DNA); *viruses* have none of these
2. viruses need a host cell to replicate in; and have no nucleus
3. a) the information given suggests larger microbe; it is more complex because it has more organelles

   b) no relationship is apparent because all microbes cause disease; more examples are given for viruses, which are the smallest
4. The same type of microbe always caused the same disease.
5. fungus
6. the bad smell made people believe that bodies were harmful; there would have been illness 'around' dead bodies because of the microbes, and so people linked it with the smell
7. a) 2000 b) 20 000
8. 0.2 μm compared to 5 μm; so the largest bacterium is 25 times bigger than the largest virus shown.

Answers to Worksheet 3.2.11

1. a) i) virus ii) bacterium iii) fungus

   b) i) cell wall, cell membrane, nucleus ii) cell wall, cell membrane iii) none
2. a) i) measles ii) HIV iii) influenza iv) tuberculosis v) athlete's foot

   b) i) *viruses*: measles, HIV, influenza; *bacteria*: tuberculosis; *fungi*: athlete's foot
3. a) i) cell wall ii) cell membrane iii) cytoplasm iv) DNA v) flagellum (tail)

   b) i) Comments such as nucleus (DNA), cell wall, cell membrane are represented.

   ii) any appropriate comments, such the cell membrane should be more flexible

2.12 Investigating the growth of bacteria

Lesson overview

Learning objectives

- Describe what bacteria need to survive.
- Investigate bacterial growth in different conditions.
- Analyse bacterial growth data.

Learning outcomes

- Describe the conditions that bacteria need to survive. [O1]
- Design and implement an investigation to compare bacterial growth in different places. [O2]
- Analyse data to make a conclusion about bacterial growth. [O3]

Skills development

- Thinking scientifically: evaluate risks
- Working scientifically: design investigations
- Learner development: ask questions

Resources needed equipment and materials as detailed in the Technician's notes; access to a camera (optional); Worksheet 3.2.12; Practical sheet 3.2.12; Technician's notes 3.2.12; Practical sheet 3.2.13 (optional); Technician's notes 3.2.13 (optional)

Digital resources Quick starter; Interactive activity: Order the surfaces based on the amount of bacteria present, from lowest to highest

Key vocabulary agar, sterile technique, inoculate, colony

Teaching and learning

Engage

- **Think, pair, share** Ask the students to **think about** what living things need to survive – they then **discuss** their ideas. Take feedback across the class. [O1]
- Ask the students to read the 'What do bacteria need to survive?' section in the Student Book. Ask pairs to **compare** the factors needed by bacteria with those needed by other living organisms. [O1]
- Students attempt task 1 of Worksheet 3.2.12. [O1]

Challenge and develop

- Show the students an agar plate and explain that agar is made from seaweed. Ask them to **respond to** verbal questions about what the agar provides for bacteria and about what conditions we may select to incubate the plates for bacteria to grow. [O1&2]
- Demonstrate inoculating an agar plate with your fingertip using the sterile technique. Ask them to **explain** why each step is important in ensuring that we grow only the type of bacteria we have chosen. Guidance is given in Technician's notes 3.2.12. [O1&2]
- The students work individually to attempt task 2 of Worksheet 3.2.12. [O1&2]

Explain

- The students work individually to **plan an investigation** to grow bacteria from different places. They should use Practical sheet 3.2.12 for guidance. [O2]

- The students use the sterile technique as far as possible and inoculate a plate with samples from four locations – these could include a sample of hair or a fingerprint. Ensure that the students check that each of their chosen locations is safe before proceeding. [O2]

  This task could be differentiated by providing lower-attaining students with locations from which to take samples. They should then make a **prediction** based on the locations provided.

- The students incubate their plates overnight and will **analyse** the results in the next lesson. [O3]

Consolidate and apply

- Ask the students to read the 'Bacteria around the home' section in the Student Book. They then work individually to answer questions 7, 8 and 9. They then pair up (preferably with partners of similar ability) and **compare** their answers. [O3]
- The students attempt task 3 of Worksheet 3.2.12.

Extend

- Tell students able to progress further that bacteria can reproduce every 20 minutes in certain conditions. Starting from one bacterium, ask them to **calculate** the number of bacteria in a culture after 5 hours. Ask the students to **suggest** why the number of bacteria does not continue to grow at this rate indefinitely. [O3]

Plenary suggestions

Learning triangle The students work individually to construct a learning triangle. They should draw a large triangle with a smaller inverted triangle that just fits inside (making four triangles). In the outer triangles, the students write something they've seen, something they've done and something they've discussed. Then, in the central triangle, they write something they've learned. [O1, 2&3]

Optional preparation for next lesson The students could **set up an experiment** in preparation for the next lesson. Instructions and notes can be found on Practical sheet 3.2.13 and Technician's notes 3.2.13.

Answers to Student Book questions

1. by breaking down food molecules to glucose (using enzymes)
2. there is no moisture present
3. low pH (acidic), body temperature, 37 °C
4. Sterile technique is used to keep equipment and conditions as sterile as possible, to avoid growth of unwanted (contaminant) bacteria.
5. to prevent contamination from other bacteria onto the plate.
6. growth of bacteria harmful to humans would be promoted
7. the toilet bowl has most bacteria followed by the kitchen cloth, pet food dish, kitchen worktop, TV remote control, and computer keyboard; comments on why there are the most in the toilet bowl because of bacteria feeding on faeces; many on the kitchen cloth because they are feeding on waste food; more on the computer keyboard because it doesn't get wiped with disinfectant as regularly as the kitchen worktop does
8. numbers will vary from house to house; and will depend on when the items were cleaned; how old the cloth is, etc.; will also vary depending on which part of the kitchen worktop is swabbed, etc.
9. which types of bacteria are present, so they know whether any are harmful to humans

Answers to Worksheet 3.2.12

1. a) nutrients b) warm c) oxygen d) 7
2. a) to remove bacteria from the hands so that these do not transfer to the plate

   b) so that the plates are not contaminated with bacteria from the loop or bud.

   c) the lid can get separated from the plate and then important information about what was grown and when would not be available

   d) to prevent contamination of the plates with bacteria from outside, and to reduce the spread of bacteria from inside the plates

   e) at 37 °C, bacteria would be grown that are suited to body temperature; and this could be a risk to humans
3. a) i) toilet floor ii) unwashed hand iii) washed hand iv) toilet bowl

   b) both have a high density of colonies, but there are more in iv) and there are at least two different types of bacteria in iv)

2.13 Understanding how antibiotics work

Lesson overview

Learning objectives

- Investigate the effect of antibiotics on bacteria.
- Explain how bacteria can become immune to antibiotics.
- Evaluate the impact of superbugs on our health.

Learning outcomes

- Investigate the effects of antibiotics on bacteria. [O1]
- Explain how bacteria become resistant to antibiotics. [O2]
- Explain 'superbugs' and evaluate their impact on society. [O3]

Skills development

- Thinking scientifically: understand how theories develop
- Working scientifically: interpret evidence
- Learner development: collaborate effectively

Resources needed agar plates from previous lesson (if prepared); mini-whiteboards; equipment and materials as detailed on the Technician's notes; Worksheet 3.2.13; Practical sheet 3.2.12; Practical sheet 3.2.13; Technician's notes 3.2.13

Digital resources Quick starter; Interactive activity: Match the bacteria-killing chemical to its purpose; Slideshow: Antibiotics - A look at how antibiotics work and how you can test the effectiveness of an antibiotic

Common misconceptions Antibiotics can treat any infection. Antibiotics can be stopped without finishing the full course without any negative effect. Overuse of antibiotics has no negative implications.

Key vocabulary disinfectant, antiseptic, antibiotic, resistance, superbug

Teaching and learning

Engage

- Ask the students to **analyse** the results from the previous lesson. They should complete their results table about the growth of bacteria from different locations. They should work individually to **analyse** and **evaluate** the investigation using Practical sheet 3.2.12 as a guide.
- Show the students a TV advert clip for a cleaning product that 'kills bacteria'. Alternatively show them an image from an advert for a cleaning product. Ask the students to **discuss** in small groups how cleaning products can kill bacteria. Take some feedback and produce a list of key words that groups share. [O1]
- Ask questions about antibiotics, letting the students answer either 'true' or 'false' using mini-whiteboards. You could use task 1 of Worksheet 3.2.13 for this activity. Do not correct any answers at this stage, but do take note of those questions causing most confusion. [O1]

Challenge and develop

- As a class, read the first and second sections in the Student Book. Ask the students to **think about** the 'true or false' questions again and to change their original answers if necessary. Ask them to display their answers again and challenge any misconceptions. [O1]
- Demonstrate how to test the effect of antibiotics on bacterial growth. If not done in the previous lesson, the students work individually to **set up an experiment** to investigate the effect of four different antibiotics on bacterial growth. Practical sheet 3.2.13 contains guidance; see also Technician's notes 3.2.13. The results can be analysed in the next lesson. If the experiment was set up in the previous lesson, the results should be analysed here. [O1]

- The students could answer questions 4, 5 and 6 in the Student Book. [O1]

Explain

- Remind the students about some of the ways in which antibiotics can kill bacteria – for example disrupt the cell wall, prevent respiration. Ask them to read the information in task 2 of Worksheet 3.2.13, which describes some of the ways in which bacteria can avoid being killed by antibiotics. [O2]
- **Pair talk/pairs to fours** Working in pairs, ask the students to **discuss** each of the avoidance methods used by bacteria. The pairs **draw** a visualisation of each of the methods. They then form fours and each pair identifies the diagram representing each of the methods. [O2]

Consolidate and apply

- **Group work** Arrange the students in small groups. Ask them to **discuss** what might happen when a bacterium with resistance to antibiotics reproduces. Ask the students to discuss whether these bacteria or other non-resistant bacteria are more likely to survive in the presence of antibiotics. Allow the groups to discuss questions 7, 8 and 9 in the Student Book. [O3]
- Ask the students to work individually to **design a leaflet** (task 3 of Worksheet 3.2.13) to explain the problem of antibiotic-resistant bacteria and the steps that people should take to avoid adding to the problem. [O3]

  This task can be differentiated by asking lower-attaining students to cover just one aspect in their leaflet – for example explain what MRSA is or how people can avoid adding to the problem.

Extend

- Ask students able to progress further to **research** MRSA in more depth. They should find out about how this multiple resistant strain developed. They should also **speculate** on the future in terms of antibiotic resistance. [O3]

Plenary suggestions

Glossary Ask the students to work individually to **produce a glossary** of key words from this lesson. The words included should those any that they regard as important and should not be restricted to those listed as 'key words' in the Student Book. Ask them to **share** some of their examples. [O1, 2&3]

Answers to Student Book questions

1. *Disinfectants* are used on surfaces and are harmful to humans. *Antiseptics* are used on the skin.
2. hospitals house bacteria on patients and surfaces, so we need both antiseptics and disinfectants
3. there was more risk of infection because bacteria were not removed from the skin before or after surgery; bacteria could enter the body
4. they disrupt the cell wall and prevent respiration
5. a) D; because this has the largest zone of clearing; this shows that antibiotic D is most effective at killing these bacteria

   b) It is not possible to tell from this information how these antibiotics affect other bacteria; it cannot be assumed that it will be as effective with different bacteria; antibiotics are specific
6. a cold is caused by a virus, so antibiotics will not treat the infection
7. Bacteria are resistant to many current antibiotics; they would not have resistance to new antibiotics, so they would be more effective
8. people in hospital or nursing homes are already weak/ill; they may have open wounds; bacteria are more prevalent in hospitals because there are more sick people than in the normal population
9. justified opinion; for example *no*, because overuse is leading to more resistant strains; or *yes*, because it is difficult to get appointments with doctors

Answers to Worksheet 3.2.13

1. a) false b) true c) true d) true e) false
2. appropriate and relevant diagrams to represent each method
3. The leaflet should contain text (plus diagrams, if chosen) making sensible suggestions such as not using antibiotics for a viral infection, completing a course of antibiotics, not stockpiling antibiotics and 'self-prescribing'.

2.14 Learning about vaccination

Lesson overview

Learning objectives

- Describe how vaccines were discovered.
- Explain how vaccines prevent a viral infection.
- Evaluate the risks involved with vaccination.

Learning outcomes

- Describe vaccines and how they were discovered. [O1]
- Explain how vaccines prevent a viral infection. [O2]
- Evaluate the risks associated with vaccination and identify bias in the data. [O3]

Skills development

- Thinking scientifically: understand how theories develop
- Working scientifically: n/a
- Learner development: plan progress

Resources needed antibiotic-sensitivity agar plates from the previous lesson (optional); photos of sufferers of smallpox and an image of the virus; large sheets of paper and pens; additional information about the history of smallpox vaccination and about how vaccines work; Worksheet 3.2.14

Digital resources Quick starter; Slideshow: Jenner, Fleming and Lister - A look at the famous scientists involved in the treatment or prevention of infectious disease; Interactive activity: Reorder the sentences to describe how a vaccination works; Hangman: Key vocabulary game

Common misconceptions Vaccines use live viruses.

Key vocabulary smallpox, vaccine, MMR

Teaching and learning

Engage

- If antibiotic-sensitivity tests were set up in the previous lesson, the results should now be collected and **analysed**. The students should use Practical sheet 3.2.13 for guidance about analysis and evaluation.
- Ask the students to look at some pictures of the effects of smallpox. Then ask them to **listen** to some background information about smallpox. Explain that smallpox appears to have been around since 10 000 BCE and that it used to kill its victims. Describe how some people tried scratching people with lesions from smallpox sufferers and that, although this gave some people resistance, the process itself killed others. Then describe the history of Edward Jenner and how he used the fluid from the much less serious cowpox to vaccinate against smallpox. [O1]
- The students should attempt task 1 of Worksheet 2.3.14. [O1]

Challenge and develop

- **Pair talk** Ask the students to read the 'How do vaccines work?' section in the Student Book. In pairs, they **explain** what they have read. Take some feedback from across the class. [O2]
- The students work individually to **draw** a cartoon strip to show how vaccines brings about immunity. They **attempt** task 2 on Worksheet 3.2.14. [O2]

  This task could be differentiated by asking lower-attaining students to produce a cartoon strip to show how Edward Jenner discovered the vaccine for smallpox.

Explain

- Ask the students to **suggest** any risks involved in vaccination. Lead a discussion about risks such as the problems of infections transferred by blood when needles were used in vaccinations and the apparent risk of autism linked with the MMR vaccine. [O3]
- The students attempt task 3 of Worksheet 3.2.14. [O3]

Consolidate and apply

- In small groups, the students **create** an advert to persuade parents to have their children vaccinated. The advert should be engaging, scientifically accurate and informative. [O2&3]

Extend

- Ask students able to progress further to **research** the influenza vaccine. They should consider why there is a need to have an influenza vaccine each year, rather than only once as for many other diseases. They should also find out about other ways of administering vaccines, other than by injection. [O1, 2&3]

Plenary suggestions

Mind map The students work in small groups to draw a mind map of the work covered on diseases and immunity. The groups then **examine** the maps produced by other groups. They then work individually to **identify** an area in which they need to make further progress. You should then use this information to support the students in their identified areas. [O1, 2&3]

Answers to Student Book questions

1. a virus
2. He took fluid from cowpox blister and inoculated a boy. Then he tested to see if the boy was immune by inoculating him with fluid from a smallpox blister. No symptoms were seen. This process was then used more widely.
3. the microbe is weakened or killed to reduce the effect, and so the symptoms are less severe
4. check that the cartoon strip is suitable and shows each stage
5. Some people are afraid of needles; some people are concerned about the spread of diseases transferred by blood (although needles are sterile nowadays).
6. opinion backed up; for example tell them that the risks are minor compared with the diseases that they protect against (e.g. mild rash compared to measles, which can cause death)
7. Concerns over the claimed risks of autism and bowel disease caused the numbers of children having the MMR vaccine to drop, which in turn caused the numbers of cases of measles, mumps and rubella to rise.
8. Provide further explanation and awareness for parents to make informed choices; use TV adverts, posters etc.

Answers to Worksheet 3.2.14

1. B; E; C; A; D
2. a) i) weakened ii) white iii) destroy iv) memory v) antibodies, symptoms

   b) suitable diagram to represent each stage
3. a) no support; there is a similar pattern in MMR and control groups

   b) Dr Wakefield might want to give the lawyers who paid him the answers that they were looking for; and they were claiming that MMR causes autism

2.15 Checking students' progress

The 'Checking your progress' section in the Student Book indicates the key ideas developed in this chapter and shows how students progress to more complex levels. It is provided to support students in:

- identifying those ideas
- developing a sense of their current level of understanding
- developing a sense of what the next steps in their learning are.

It is designed either to be used at the end of a chapter to support an overall view of the progress, or alternatively during the teaching of the unit. Students can self assess or peer assess using this as a basis.

It would be helpful if students can be encouraged to provide evidence from their understanding or their notes to support their judgements. In some cases it may be useful to explore the difference in the descriptors for a particular idea so that students can see what makes for a 'higher outcome'.

It may be useful with some descriptors to provide examples from the specific work done, such as an experiment undertaken or an explanation developed and recorded. If marking and feedback uses similar ideas and phrases, this will enable students to relate specific marking to a more general sense of progress.

To make good progress in understanding science students need to focus on these ideas and skills:

Students who are making modest progress will be able to:	**Students who are making good progress will be able to:**	**Students who are making excellent progress will be able to:**
Give examples of some different types of drugs.	Describe the effects of different types of drugs on the body.	Explain the effects of different drugs on the body, including harmful effects.
Identify parts of the body damaged by smoking.	Explain how the body is damaged by smoking and by passive smoking.	Examine data about smoking and cancer, and draw a conclusion about the correlation.
Describe the effects of cannabis on the body.	Suggest some reasons why people use cannabis.	Present, using evidence, arguments for and against the legalisation of cannabis.
Describe some effects of alcohol on the body.	Describe and explain several effects of alcohol on the body.	Suggest how alcoholism affects society.
Describe what addiction is and give examples.	Describe the changes in the brain caused by drugs and how this can lead to addiction.	Explain the effects of withdrawal from drug dependency on the body, including the brain.
Describe and give examples of a way in which diseases are spread.	Describe several examples of how specific diseases are spread and suggest how their spread may be reduced.	Consider suggestions to reduce the spread of specific diseases and justify decisions.
Describe the body's mechanisms to prevent infection.	Describe the roles of white blood cells in fighting infection.	Explain why we rarely catch the same infectious disease twice, but may catch influenza over and over again.
State examples of diseases caused by microbes.	Describe the characteristics of different microbes.	Evaluate a model of a type of microbe.
Describe the conditions that bacteria need to survive.	Compare bacterial growth in different parts of the home.	Analyse data about bacterial growth.
Describe the effect of antibiotics on bacteria.	Explain how bacteria become resistant to antibiotics.	Explain what superbugs are and evaluate their impact on society.
Describe what a vaccine is and how vaccines were discovered.	Explain how vaccines prevent a viral infection.	Evaluate the risks associated with vaccination.

2.16 Answers to Student Book Questions

This table provides answers to the Questions section at the end of Chapter 2 of the Student Book. It also shows how different questions assess attainment in terms of the focus and style of a question as well as the context. Question level analysis can indicate students' proficiency in approaching different aspects of scientific understanding and different types of answer.

Q	Answer		Focus			Style			Context		
		Marks available	Knowledge & understanding	Application	Evaluation of evidence	Objective test question	Short written answer	Longer written answer	Drugs	Disease	Preventing and treating infection
1	c	1	x			x			x		
2	b	1	x			x			x		
3	a	1	x			x				x	
4	d	1	x			x					x
5	disinfectants kill microbes on surfaces/inanimate objects whereas antiseptics kill microbes on skin OR accept a comparison of strength (disinfectants stronger, antiseptics weaker) do not award marks for simply saying that they kill microbes, because both do	1 1	x x				x x				x x
6	any **two** from: • cell wall • cell membrane • circular DNA	2	x				x			x	
7	any **four** from: • physical dependency on a drug • addict needs to take the drug to feel normal • may suffer withdrawal symptoms if don't take the drug • example of withdrawal symptoms • example of addictive drugs such as alcohol, nicotine, heroin • mention of genetic factor in vulnerability to addiction	4	x					x	x		
8	a	1		x		x			x		
9	b	1		x		x			x		
10	a	1		x		x					x

Q	Answer	Marks available	Focus: Knowledge & understanding	Focus: Application	Focus: Evaluation of evidence	Style: Objective test question	Style: Short written answer	Style: Longer written answer	Context: Drugs	Context: Disease	Context: Preventing and treating infection
11	any **two** from: • cover mouth when sneezing/ coughing • spread children around the school • turn off air conditioning	2		x			x			x	
12	d	1		x		x				x	
13	number of cases should decrease	1		x			x		x		
	Any **one** from: • passive smoking passes chemicals from cigarettes to non-smokers • cars are closed spaces where cigarette chemicals are passed to non-smokers easily	1		x			x		x		
14	recommend D	1		x				x			x
	any **three** from: • D; has biggest zone of clearing/ kills most bacteria • C/ B; kills fewer bacteria/ smaller zone of clearing • A; kills no bacteria because there is no zone of clearing • refer to the way the handwash kills bacteria; e.g. antiseptics cause cells to burst	3		x				x			x
15	virus; reasons such as protein coat, strand of DNA, no cell wall, no cell membrane	1			x		x			x	
	antibiotics not useful (only kill bacteria)	1			x		x			x	
16	when antibiotics used; only those with resistance survive	2			x			x			x
	these bacteria reproduce and produce populations of resistant bacteria	1			x			x			x
	tested using antibiotic discs soaked in the antibiotic on a lawn of bacteria at various times (or other suitable method)	1			x			x			x
	Total possible:	**30**	**12**	**12**	**6**	**8**	**10**	**12**	**10**	**8**	**12**

3.1 Introduction

When and how to use these pages

The Introduction in the Student Book indicates some of the ideas and skills in this topic area that students will already have met from KS2 or from previous KS3 work, and provides an indication of what they will be studying in this chapter. *Ideas you have met before* is not intended to comprehensively summarise of all the prior ideas, but rather to point out a few of the key ones and to support the view that scientific understanding is progressive. Even though students might be meeting contexts that are new to them, they can often use existing ideas to start to make sense of them.

In this chapter you will find out indicates some of the new ideas that the chapter will introduce. Again, it isn't a detailed summary of content or even an index page. Its purpose is more to act as a 'trailer' and generate some interest.

The outcomes, then, will be recognition of prior learning that can be built on, and interest in finding out more.

There are a number of ways this can be used. You might, for example:

- Use *Ideas you have met before* as the basis for a revision lesson as you start the first new topic.
- Use *Ideas you have met before* as the centre of spider diagrams, to which students can add examples, experiments they might have done previously or what they found interesting.
- Make a note of any unfamiliar/difficult terms and return to these in the relevant lessons.
- Use *In this chapter you will find out* to ask students questions such as:
  - Why is this important?
  - How could it be used?
  - What might we be doing in this topic?

Overview of the chapter

In this chapter, students will learn about how metals are obtained from the Earth. They will consider the environmental impacts of mining and how these may be mitigated. The students will apply their knowledge and understanding of the reactivity series and explain what happens in displacement reactions. They will link ideas about reactivity to the ease with which metal carbonates are decomposed. Finally, these ideas will be applied to the extraction of less reactive metals with carbon. The students will develop in their abilities to write word and balanced symbol equations, and apply ideas about conservation of mass to the extraction of metal ores. The students will consider energy changes in chemical reactions using an understanding of bond-making and bond-breaking. They will explain why exothermic and endothermic changes take place and how these can be usefully applied. They will be introduced to the idea of catalysts and how they can change the rate of chemical reactions. Finally the students will learn about the structure, properties and uses of new materials, including ceramics, polymers and composites. They will consider the importance of these materials and what they can offer in place of metals.

The students will be offered many opportunities to plan investigations, to make detailed observations during investigations involving displacement reactions and reactions in which there are temperature changes and catalysts. They will also present and interpret data in the form of tables and graphs.

Obstacles to learning

Students may need extra guidance with the following terms and concepts:

- **Ores** Some students may think of ores simply as rocks and minerals. The idea that an ore is a rock that contains enough of the metal to make extraction worthwhile should be emphasised.
- **Thermal decomposition** Students may be unaware that not all metal carbonates can be thermally decomposed easily. They may believe that compounds with similar groups behave in the same way. By linking the compounds to the reactivity of the metals, the students will begin to develop the idea that patterns of reactivity exist.
- **Displacement** Students may think that all metals react in the same way and have difficulty with the idea that some are more reactive than others. The use of models and role-play will help to clarify these ideas.

- **Exothermic and endothermic reactions** Students may be familiar with the idea that energy is given out in reactions, resulting in an increase in temperature. However, the idea that some reactions take in energy causing the temperature to fall may cause cognitive conflict. They may think that because energy is being taken in (during an endothermic change) it should cause the reactants to become hotter, and so the temperature should rise. In fact the balance between the energy taken in to break reactant bonds and the energy given out means that, overall, more energy is removed from the surroundings and the temperature decreases.
- **Catalysts** Many students will automatically think of a catalyst as a reactant. Recovering a catalyst and reusing it several times will help to show that catalysts do not get used up in reactions.
- **Ceramics, polymers and composites** Students are likely to think that all these materials are human-made – they will be introduced to the natural examples in the Student Book to understand how they are structured.

	Topic title:	**Overarching objectives:**
2	Obtaining metal from ores	Earth as a source of limited resources
3	Understanding reactivity	The order of metals and carbon in the reactivity series; representing chemical reactions using formulas and using equations; thermal decomposition
4	Making use of displacement reactions	The order of metals and carbon in the reactivity series; representing chemical reactions using formulas and using equations; displacement reactions; conservation of mass, changes of state and chemical reactions
5	Using carbon to extract iron	The use of carbon in obtaining metals from metal oxides; representing chemical reactions using formulas and using equations; conservation of mass, changes of state and chemical reactions
6	Extracting copper, lead and zinc	The use of carbon in obtaining metals from metal oxides; representing chemical reactions using formulas and using equations
7	Looking at the impact of metal extraction	The use of carbon in obtaining metals from metal oxides; the production of carbon dioxide by human activity and the impact on climate; Earth as a source of limited resources and the efficacy of recycling
9	Understanding exothermic reactions	Internal energy stored in materials; exothermic chemical reactions (qualitative); comparing the starting with the final conditions of a system and describing increases and decreases in the amounts of energy associated with chemical compositions
10	Comparing endothermic and exothermic changes	Exothermic and endothermic chemical reactions (qualitative); comparing the starting with the final conditions of a system and describing increases and decreases in the amounts of energy associated with chemical compositions
11	Explaining the use of catalysts	What catalysts do
12	Exploring ceramics and their properties	Properties of ceramics (qualitative)
13	Matching properties of ceramics to their uses	
14	Exploring natural polymers	Properties of polymers (qualitative)
15	Using human-made polymers	
16	Explaining composites	Properties of composites (qualitative)
17	Using human-made composites	

3.2 Obtaining metals from ores

Lesson overview

Learning objectives

- Recognise how abundant common ores are in the Earth.
- Explain how ores are extracted from the Earth.

Learning outcomes

- Describe some methods to extract common ores. [O1]
- Analyse data to describe the abundance of different metal ores; describe methods of mining ores from the Earth. [O2]
- Compare and contrast different extraction methods. [O3]

Skills development

- Thinking scientifically: ask questions
- Working scientifically: interpret evidence
- Learner development: communicate effectively

Resources needed sticky notes; a world map; Worksheet 3.3.2

Digital resources Quick starter; Interactive activity: Drag the examples of substances to the correct group - native metals, non-native metals or ores; Video; Interactive activity: Place, in order, the metals from most to least reactive

Common misconceptions All metals are native; ores are just rocks.

Key vocabulary ore, mineral, smelting

Teaching and learning

Engage

- Introduce the term 'ore' using some real examples if possible, or use the images from the Student Book. [O1]
- Ask the students to **devise** some questions on sticky notes that they would like to ask about metal ores – they will come back to these later in the lesson. [O1&2]

Challenge and develop

- Ask the students if it is possible to **predict** which metal comes from which ore, just by looking at it. Show them the chemical formulas of some common ores. Explain that these formulas are not always accurate because other minerals can be mixed up with the ore. Revise ideas about physical and chemical changes – establish that both are needed in separating a metal from an ore. [O1]
- Discuss the historical discovery of metals using ideas from the Student Book. If possible show a video of early smelting of copper, lead and/or tin. Ask the students to answer questions 1–4 in the Student Book. [O1&2]
- If possible, show videos of surface mining and deep mining of different metal ores. Discuss why different methods are used, using ideas about volcanic and tectonic plate activity to explain why seams of ore are found at different depths in the Earth's crust – avoid too much detail. Use the Student Book to help. [O1&2]
- Explore the differences between surface mining and deep mining. [O3]

Explain

- Ask the students to complete task 2 of Worksheet 3.3.2, about a bar graph of abundances of elements (Figure 3.3.2c of the Student Book). They should **explain** why the data has been presented in this format and

explore other ways in which it could be presented. Discuss the advantages and disadvantages of their suggestions. Discuss what the data is showing and how accurate it is likely to be. [O2]

Consolidate and apply

- **Small groups** Ask the students to complete task 1 of Worksheet 3.3.2, **designing** a 'trump card' game about ores. [O1, 2&3]

  Higher-attaining students should do further **research** of ores to enhance their games.

- Select different students to **present** what they have found out about their ores. Using a map of the world, ask the students to **locate** where particular ores are found. [O1, 2&3]

  Higher-attaining students should answer Student Book questions 5–7. [O3]

Extend

- Ask students able to progress further to research the process of electrolysis and **produce a poster** about it, **explaining** in simple terms how it works and which metals are extracted using this process. They could **evaluate** the strengths and weaknesses of electrolysis. [O3]

Plenary suggestions

Questions-and-answers game The students use the sticky note questions they developed at the start of the lesson. See how many questions they can answer in two minutes. [O1, 2&3]

Write sentences The students write three sentences using the key words for the topic. [O1, 2&3]

Answers to Student Book questions

1. Ores are chemical compounds containing a metal. Because the metal in an ore is chemically combined with other elements; this gives the ore different properties to the metal they contain.
2. Gold is very unreactive and does not combine with other elements easily.
3. Metals that are uncombined or easy to extract from their ores were discovered by chance much earlier in history. Metals that are harder to extract from their ores were discovered only when chemical methods became more sophisticated. Most ores do not look like their metals.
4. Aluminium is very hard to separate it from its ore; the ore looks like an 'ordinary rock' and it is hard to imagine you can extract a metal from it.
5. electricity has to be used; this was not possible until the late 1800s
6. reactive metals form much more stable compounds; these are harder to break down than the compounds formed from unreactive metals
7. There are huge problems with water pressure and safety issues when drilling into the ocean floors. It is a very expensive process and it is difficult to remove the waste material. Damage to the local environment caused by drilling, disturbing the ocean bed and creating lots of suspended 'dust' is likely to affect organisms living in the area.

Answers to Worksheet 3.3.2

1. Examples of information on a trump card for an ore: Name / How many countries is it found in? / How many main uses does it have? / How many tonnes are mined? / How interesting is the other information?
2. a) a pie chart could be made using the data

   b) Students should be encouraged to find as much information from the bar graph as possible. Here are some suggested answers: there is nearly twice as much oxygen compared to silicon; oxygen is the most abundant element; magnesium is the least abundant; there is twice as much aluminium compared to calcium; there is nearly three times as much silicon compared to aluminium.
3. a) Deep mining: *advantages* – not as much waste rock produced; less visible impact so easier to conceal; *disadvantages* – more risk of subsidence; heavy machinery required to cut the rock; greater risk of cave-ins; harder to remove waste rock.

   Surface mining: *advantages* – less risk of subsidence; *disadvantages* – more scarring on the landscape; more waste rock produced.

   b) Electrolysis uses electricity to separate the metal from other elements in the ore; it is only needed for extracting elements that are more reactive than carbon. Using electricity can be expensive, and relies on the use of fossil fuels to generate electricity, and so contributes to global warming. Use of coke is much cheaper than electricity, although carbon dioxide is made in the process, which contributes to global warming.

3.3 Understanding reactivity

Lesson overview

Learning objectives

- Use evidence to identify the reactivity series of metals.
- Represent reactions using formulas and equations.

Learning outcomes

- Write word equations to represent the decomposition of metal carbonates; use observation to make inferences about the reactivity of different metals. [O1]
- Use data to make inferences about metal reactivity. [O2]
- Write balanced symbol equations for the decomposition of metal carbonates; explain how decomposition of metal carbonates relates to metal reactivity. [O3]

Skills development

- Thinking scientifically: use units and nomenclature
- Working scientifically: interpret evidence
- Learner development: collaborate effectively

Resources needed equipment and materials as detailed in the Technician's notes; Worksheet 3.3.3; Practical sheet 3.3.3; Technician's notes 3.3.3

Digital resources Quick starter; Interactive activity: Place, in order, the metals from most to least reactive; Slideshow: Reactivity of metals - some examples

Common misconceptions All chemical compounds are stable and cannot be broken down using heat.

Key vocabulary reactivity, metal carbonate, thermal decomposition

Teaching and learning

Engage

- Demonstrate how sodium, iron and copper behave when added to water (see Technician's notes 3.3.3), or use the images from the Student Book. Ask the students to **identify** key differences between the reactions and to **suggest** some reasons for these. Establish that some metals are more reactive than others. Ask them to **order** the metals sodium, iron and copper according to their reactivity from their observations. [O1]
- Use the 'Signs of how reactive metals are' section in the Student Book to identify other signs of how metals differ in reactivity and ask the students to answer questions 1–3. [O1]

Challenge and develop

- Show the students the formulas of some metal ores. Ask them to **identify** which ores commonly occur as the metal carbonate. Explain that heat can be used to decompose the metal carbonate and break it down. Ask them to **predict** what the products of thermal decomposition might be and to write a word equation for the thermal decomposition of a given metal carbonate. [O1]
- Demonstrate the test for carbon dioxide (turning limewater milky) and ask the students to **explain the link** between this and the decomposition of metal carbonates. O1]
- **Group work** Arrange the students into small groups. Ask them to follow the instructions on Practical sheet 3.3.3 to **make** and **record observations** about how easy or hard it is for metal carbonates to be decomposed. They should **test** any gases for the presence of carbon dioxide. Refer to Technician's notes 3.3.3 for important safety considerations. [O1]
- The students should **write word equations** for any reactions observed during their practical work. [O1]

  Higher-attaining students can attempt to **write balanced symbol equations** for the reactions. [O3]

Explain

- Ask the students, still in groups, to **prepare a presentation** to **share** and **explain** their results. Select different groups to share their observations and explanations. They should attempt to **make links** between the ease of decomposition of the metal carbonate and the reactivity of the metal. [O2]

  Higher- attaining students should use **role-play**, or **annotated diagrams**, and the information from the Student Book to **explain** why some metal carbonates are easier to decompose than others. [O3]

Consolidate and apply

- The students should complete the appropriate tasks of Worksheet 3.3.3. [O1, 2&3]
- Ask them to answer questions 4 and 5 in the Student Book. [O2]

  Ask higher-attaining students to answer questions 6 and 7. [O3]

Extend

- Ask students able to progress further to **research** other reactions in which the reactivity of a metal plays an important part. They could **produce a short presentation** to the class about the importance of differences in reactivity between different metals. [O3]

Plenary suggestions

Make a summary The students should **summarise** the importance of metal ore decomposition. [O1, 2&3]

Learning triangle They should **complete** a learning triangle to review what they have learned in this lesson. [O1, 2&3]

Answers to Student Book questions

1. sodium > magnesium > copper
2. it will react less vigorously than sodium, so it is unlikely to be explosive; but it will be more vigorous than iron, so it will bubble a lot and may get warm
3. zinc + hydrochloric acid → zinc chloride + hydrogen

   potassium + hydrochloric acid → potassium chloride + hydrogen

   gold + hydrochloric acid (no reaction)
4. magnesium > calcium > strontium > barium
5. metals that need the least amount of energy to decompose their metal carbonate; less reactive metals like copper
6. Potassium carbonate; it is the most reactive metal and would have the most stable carbonate.
7. $Na_2CO_3 \rightarrow Na_2O + CO_2$

Answers to Worksheet 3.3.3

1. sodium would react very violently; magnesium would bubble vigorously; iron would show a very slow reaction. More heat would be produced in the sodium reaction; then magnesium; and only a little from iron
2. a) metals displaced: i) zarkon ii) none iii) zarkon iv) carkon v) carkon

   b) order of reactivity: darkon > arkon > carkon > zarkon
3. a) i) $Al_2(CO_3)_3 \rightarrow Al_2O_3 + 3CO_2$

   ii) $K_2CO_3 \rightarrow K_2O + CO_2$

   b) i) $CuCO_3 \rightarrow CuO + CO_2$

   ii) $CaCO_3 \rightarrow CaO + CO_2$

   iii) $B_2(CO_3)_3 \rightarrow B_2O_3 + 3CO_2$

3.4 Making use of displacement reactions

Lesson overview

Learning objectives

- Represent and explain displacement reactions using formulas and equations.
- Make inferences about reactivity from displacement reactions.

Learning outcomes

- Write word equations to represent displacement reactions; give some uses for displacement reactions. [O1]
- Use models to explain displacement and relate it to the reactivity series. [O2]
- Write balanced symbol equations for displacement reactions. [O3]

Skills development

- Thinking scientifically: use nomenclature
- Working scientifically: interpret evidence
- Learner development: collaborate effectively

Resources needed a previously prepared iron nail that has been in a solution of copper sulfate overnight; iron nail and freshly prepared copper sulfate solution; equipment and materials as detailed in the Technician's notes; Worksheet 3.3.4 (page 2 copied onto cards); Practical sheet 3.3.4; Technician's notes 3.3.4

Digital resources Quick starter; Interactive activity: Match the reactants to the products; Interactive activity: Place, in order, the elements from most to least reactive

Common misconceptions All metals react in the same way with same degree of reactivity.

Key vocabulary displacement, reactivity series

Teaching and learning

Engage

- Demonstrate putting an iron nail in a solution of copper sulfate – show a previously prepared sample of an iron nail in copper sulfate that has been left overnight. Ask the students to **make observations**. Write down the reactants and ask the students to predict what the products might be. Alternatively use the images in the Student Book. [O1]
- Tell the students that you will repeat the reaction using a strip of magnesium instead of iron. Ask them to **recap** what they learned from last lesson and to **predict** how this reaction will be different to that of the iron **giving reasons**. Ask them to **write word equations** to predict the products. [O1]

Challenge and develop

- Select three students to **demonstrate a role-play** of displacement reactions, using word cards of the reactants as given on Worksheet 3.3.4. Ask them to role-play the reaction between iron and copper sulfate. Emphasise that, because iron is more reactive than copper, it will 'snatch away' the sulfate, which is bonded to the copper. Use the students to demonstrate how this works, with the student holding the iron card 'snatching away' the student holding the sulfate card. [O1&2]
- Introduce and define the term 'displacement reaction'. Use the 'Using displacement reactions' section in the Student Book to discuss the importance of displacement reactions and how they can be useful. [O1]
- **Group work** Students, in small groups, should **follow the instructions** on Practical sheet 3.3.4. They should **work out** the order of reactivity of the different metals by reacting them with solutions of different metals. [O1]

Explain

- Provide students (in groups of four) with the word cards from Worksheet 3.3.4 and the reactivity series shown in Figure 3.3.4c in the Student Book. Ask them to use role-play to **explain** what will happen in the reactions they have just investigated. Select different students to **demonstrate** the reactions. [O1&2]

  Higher-attaining students should **write balanced symbol equations** for all equations. [O3]

Consolidate and apply

- Ask the students to use the list of reactivity of metals from the Student Book and the cards from Worksheet 3.3.4 to **devise** five reactions of their own to show the idea of displacement. [O1, 2&3]
- They should complete tasks 1 and 2 of Worksheet 3.3.4, **writing word equations**, using the card sort first to work out what is happening. [O1&2]

  Higher-attaining students should complete task 3 of the worksheet and use the symbol sheet to **write balanced symbol equations** of their own. [O3]
- The students should answer all the appropriate questions from the Student Book . [O1, 2&3]

Extend

- Ask students able to progress further to carry out some **research** and **make a slideshow** on three important displacement reactions, giving word and balanced symbol equations for all reactions they identify. [O3]

Plenary suggestions

Make a poster The students could **prepare a poster** for another class to **explain** what displacement reactions are and how they are useful. Word and balanced symbol equations should be included. [O1, 2&3]

Other models The students could think of other models/analogies to explain the displacement process. [O2]

Answers to Student Book questions

1. magnesium
2. iron sulfate + magnesium → magnesium sulfate + iron
3. copper will displace the silver; to make copper nitrate and silver
4. $Cu + 2Ag\ NO_3 \rightarrow Cu(NO_3)_2 + 2Ag$
5. chlorine > bromine > iodine
6. $K + AgNO_3 \rightarrow KNO_3 + Ag$ (gold and platinum are very unreactive and make few compounds)
7. Copper is less reactive than hydrogen and will not displace it.
8. sodium is an expensive metal, so it is not commercially viable to do this

Answers to Worksheet 3.3.4

1. a) the product magnesium chloride is wrong; it should be magnesium sulfate

   b) this implies no reaction, which is wrong because iron is more reactive than lead; the products should be iron nitrate + lead

   c) this shows that a reaction has taken place, which is wrong; in fact no displacement occurs and the reactants remain unchanged; (however, if the potassium chloride was dissolved in water, the sodium would react vigorously making hydrogen)

   d) the product iron is wrong; it should be copper
2. a) correct word equations with the same elements in the reactants as in the products; in the products the more reactive metal should be combined with the radical of the less reactive metal

   b) i) tin will not react with iron sulfate ii) tin will react with lead sulfate to make lead and tin sulfate
3. correctly balanced symbol equations with the same number of atoms of each element on either side of the equation; the formulas of the reactants and products must not be altered

   e.g. $2Al + 3CuSO_4 \rightarrow Al_2(SO_4)_3 + 3Cu$

3.5 Using carbon to extract iron

Lesson overview

Learning objectives

- Represent displacement reactions with carbon, metal oxides and iron using formulas and equations.
- Explain how mass is conserved in the extraction of metals.

Learning outcomes

- Describe the reaction between carbon and iron oxide; use the reactivity series to explain why carbon is used to extract metals. [O1]
- Describe the process for extracting iron from the blast furnace. [O2]
- Use balanced symbol equations to make predictions about the mass of iron produced, showing that mass is conserved; explain the advantages of using carbon to extract metals. [O3]

Skills development

- Thinking scientifically: use nomenclature
- Working scientifically: develop explanations
- Learner development: collaborate effectively

Resources needed two or three samples of haematite; equipment and materials as detailed in the Technician's notes; Worksheet 3.3.5; Technician's notes 3.3.5

Digital resources Quick starter; Interactive activity: Drag the statements to the correct group; Slideshow: The blast furnace - What goes in, and what comes out

Common misconceptions Carbon is not part of the reactivity series.

Key vocabulary coke, reduction, oxidation, blast furnace

Teaching and learning

Engage

- Show the students samples of haematite, if possible, and discuss its chemistry. Alternatively use images from the Student Book. [O1]
- Show a short video clip of early people making iron using charcoal, or use the Student Book. Discuss what is happening in chemical terms when iron ore is heated using carbon. Establish that this is a displacement reaction and introduce the term 'reduction' to mean the removal of oxygen from an oxide. [O1]

Challenge and develop

- Demonstrate the production, on a small scale, of iron from iron oxide using carbon, following the procedure given in Technician's notes 3.3.5. Arrange the students in small groups. They should **write word equations** to **summarise** the reaction. [O1]

  Higher-attaining students should **write balanced symbol equations**. [O3]
- Show a video, if possible, of iron being extracted in a blast furnace. Discuss how this process works to extract the iron. Alternatively use the Student Book to explain how this reaction is completed on a large scale. [O1&2]

Explain

- **Group talk** The students, still in their groups, should **discuss** their observations.They should **explain** what happened to the mass of the reactants and products before and after the reaction. [O1&3]
- Show some different samples of iron ore – for example magnetite if available and other samples of haematite. Discuss how much iron each sample contains. Establish that different samples will contain different amounts of iron. Use the 'Predicting the mass of iron' section in the Student Book to explain how to predict the mass of iron knowing its purity. [O3]

Consolidate and apply

- The students should answer all the Student Book questions and complete the tasks of Worksheet 3.3.5. [O1, 2&3]
- **Pair work** Ask the students to **discuss** different ways they could make the ancient process of extracting iron from its ore more efficient. [O1, 2&3]

Extend

- Ask students able to progress further to **carry out research** about the blast furnace in more detail. They could **draw** an annotated diagram **explaining** each of the stages, to show how the process maximises the production of iron. [O3]

Plenary suggestions

Writing definitions The students could **write** their own definitions of the key words from the Student Book. [O1&2]

Drawing a flow chart They could **draw** a flow chart to **explain** how iron is made from iron ore, including all key terms.

Higher-attaining students should **explain** how the maximum mass of iron is obtained from its ore. [O1, 2&3]

Answers to Student Book questions

1. carbon would eventually displace the zinc from zinc sulfide, to form zinc metal, because carbon is more reactive
2. carbon + lead oxide → carbon dioxide + lead
3. so that the reactions can occur at a faster rate; powdered iron and carbon have a higher surface area than lumps of carbon and iron
4. this helps the carbon to remove the oxygen from the iron more effectively; higher temperatures speed up the reaction
5. $Fe_3O_4 + 2C \rightarrow 2CO_2 + 3Fe$
6. theoretical amount = 250 × 0.3 = 75 tonnes

Answers to Worksheet 3.3.5

1. a) reactions occur with ii), iv), v).
   b) ii); iv); v)
   c) copper oxide + carbon → copper + carbon dioxide
   iron oxide + carbon → iron + carbon dioxide
   lead oxide + carbon → lead + carbon dioxide
2. a) coke and iron ore (iron oxide)
   b) molten iron
   c) carbon dioxide
   d) iron oxide + carbon → iron + carbon dioxide
3. *Advantages* of using more reactive metals: faster reactions occur
   Disadvantages: expensive because reactive metals are expensive to extract from their ores
   Advantages of using carbon: it is cheap
   Disadvantages: carbon dioxide is made as a product; this is a greenhouse gas

3.6 Extracting copper, lead and zinc

Lesson overview

Learning objectives

- Explain how copper, lead and zinc are extracted from their ores.
- Calculate the yield of the extraction process.

Learning outcomes

- Recognise why carbon is needed in the extraction of lead, copper and zinc; write word equations for the reactions between carbon and metal ores. [O1]
- Describe the extraction processes for lead, copper and zinc. [O2]
- Use balanced symbol equations to make predictions about the mass of lead, copper or zinc produced; work out the efficiency of the extraction process. [O3]

Skills development

- Thinking scientifically: evaluate risks
- Working scientifically: record evidence
- Learner development: collaborate effectively

Resources needed samples of galena, malachite and/or sphalerite; equipment and materials as detailed in the Technician's notes; Worksheet 3.3.6; Practical sheet 3.3.6, Technician's notes 3.3.6

Digital resources Quick starter; Interactive activity: Drag the statements to the correct group - reduction, oxidation or decomposition

Common misconceptions Carbon is only used to extract iron.

Key vocabulary yield

Teaching and learning

Engage

- **Pair work** Working in pairs, the students play 'Pictionary' to **recap key vocabulary**. Use the following words: 'carbon', 'reduction', 'displacement', 'ore', 'reactivity series'. [O1]
- Show them the reactivity series in the Student Book (Figure 3.3.4c). Ask them to **predict** which other metals might be extracted by using carbon. [O1]

Challenge and develop

- Show the students samples of lead ore (galena), zinc ore (sphalerite) and copper ore (malachite), or use images from the Student Book. Discuss the chemical composition of the ores and what needs to be done to extract the metal. If possible, show a video of the extraction of lead, copper and/or zinc metals from their ores. Use the Student Book to explore why sulfide ores need to be first roasted in oxygen to convert them to the oxide, before carbon can be used to reduce them. [O1]
- Demonstrate the extraction of lead from its oxide, using the procedure given in Technician's notes 3.3.6. Discuss the safety requirements of the extraction process. [O1&2]
- **Group work** Arrange the students into small groups. They should **follow the procedure** on Practical sheet 3.3.6 to make copper from copper oxide. They should first examine the procedure and **identify** and **evaluate the risks** involved, and decide how to minimise the risks in the investigation. [O1&2]

  Higher-attaining students should **write balanced symbol equations** for the reaction and **predict** the amount of copper to be made. They should **record** their observations, including the mass of the reactants before and the products after. They should **account** for any apparent losses in mass. [O3]

Explain

- **Pair work** Ask the students to **explain** their observations, along with word and symbol equations, accounting for any apparent losses in mass. Select different pairs to **share** their explanations. [O1, 2&3]

  Higher-attaining students could **calculate** the efficiency of their process, given an appropprimate percentage purity. [O3]

Consolidate and apply

- Ask the students to use the Student Book to **compare and contrast** the methods of extraction for copper, lead and zinc. They could draw up a table of similarities and differences. [O1&2]
- They should answer all the appropriate questions in the Student Book and complete the tasks of Worksheet 3.3.6, according to their abilities. [O1, 2&3]

Extend

- Ask students able to progress further to **research** how metals higher than carbon in the reactivity series are extracted.

Plenary suggestions

Just a minute The students could be asked to **present** a one-minute talk on the extraction of metals using carbon. [O1, 2&3]

Hot seat Questions could be posed to a student about the extraction processes of copper, zinc and lead from their ores. [O1, 2&3]

Answers to Student Book questions

1. carbon dioxide
2. copper oxide
3. the steps in a suitable flow chart are:

   lead sulfide ore is crushed into small particles; oil and water are added; air is blown upwards through mixture; lead sulfide is separated; the lead sulfide is roasted; the lead oxide is roasted with carbon; lead is made
4. zinc sulfide + oxygen → zinc oxide + sulfur dioxide

   zinc oxide + carbon → zinc + carbon dioxide
5. Yield = 500 ÷ 800 × 100 % = 62.5 %
6. 50% of 446 = 223 tonnes of lead oxide are in the ore; 200 ÷ 223 × 100 % = 90 % yield

Answers to Worksheet 3.3.6

1. b, d

   copper oxide is being reduced in b), because carbon is removing oxygen from it

   iron oxide is being reduced in d), because carbon is removing oxygen from it
2. a) tin oxide + carbon → tin + carbon dioxide

   b) The tin sulfide is roasted in oxygen to convert it to tin oxide.

   c) With *stannite* the ore needs to be roasted in oxygen, and then froth flotation used to concentrate the tin oxide made and remove it from the impurities in the ore. Sulfur dioxide, which contributes to acid rain, is made in the process.

   With *cassiterite* the crushed ore is just roasted with carbon to extract the metal.
3. a) $SnS + 2O_2 \rightarrow SnO_2 + SO_2$

   $SnO_2 + C \rightarrow Sn + CO_2$

   b) 750 × 0.47 = 352.5 tonnes

   c) 140 ÷ 352.5 × 100 = 40%

3.7 Looking at the impact of metal extraction

Lesson overview

Learning objectives

- Describe the environmental impacts of metal extraction.
- Describe how recycling of metals reduces damage to the environment.

Learning outcomes

- Describe the main issues concerning metal mining and extraction processes; describe the main solutions to minimise impacts. [O1]
- Explain how metals are recycled and how this affects the environment. [O2]
- Evaluate the positive and negative aspects of metal mining and extraction. [O3]

Skills development

- Thinking scientifically: ask questions
- Working scientifically: develop explanations
- Learner development: communicate effectively

Resources needed sticky notes; Worksheet 3.3.7 (with page 2 copied onto card)

Digital resources Quick starter; Video; Interactive activity: Match the pollution to its effect; Slideshow: Recycling metals - A look at how we recycle metals, including e-waste; Hangman: Key vocabulary game

Common misconceptions Environmental impacts from mining are just visual.

Key vocabulary reclamation, subsidence, leachate, recycling

Teaching and learning

Engage

- Ask the students to work in pairs to **write a list** of all the different environmental impacts associated with the mining and extraction of metals. Take feedback from selected pairs and write their ideas on the board. [O1]
- The students should **devise** as many questions as they can about the environmental impacts of mining and metal extraction and write them on sticky notes. [O1]

Challenge and develop

- Show a video about the impacts of mining and/or use the Student Book. Discuss other issues associated with mining and extraction that are not already covered. Include less obvious impacts like the release of carbon dioxide from heavy machinery and the contribution to acid rain from roasting metal sulfide ores. [O1]
- **Pair work** Ask the students to **create** a poster to **summarise** the impacts of mining and the damage it can potentially cause, as indicated in task 1 of Worksheet 3.3.7. They should use the information on the cards from the worksheet to help them. Ask them to **peer-assess** each other's posters, **identifying** two areas of strength and one area for development. Allow improvement time for students to **respond** to comments. [O1]
- Discuss methods of mitigating the impacts of mining and extraction, including the recycling of metals. Ask the students to answer the questions in task 2 of Worksheet 3.3.7. They should add any further ideas about recycling to their posters. [O2]

Explain

- Select different students to show their posters, **explaining** how recycling and other methods can be used to mitigate the effects of mining and metal extraction. [O2]

  Higher-attaining students should **summarise** the positive and negative effects of mining and metal extraction and **justify** whether or not we should continue to carry on using metals. [O3]

Consolidate and apply

- **Group work** Arrange the students in groups of four. Ask them to use the Student Book and the information from Worksheet 3.3.7. This has a scenario involving four different people involved with different agendas. Ask students to **role-play** being one of the four people given. They should use the scenario cards to **discuss** what should be done, using their posters and/or any evidence they have to **support their arguments**. Select some groups to share their ideas and solutions. [O1, 2&3]
- The students should answer the questions from the Student Book. [O1, 2&3]

Extend

- Ask students able to progress further to **write a detailed argument** for the environmental officer and the mining manager in the role-play. They **present** their ideas as a summary of the arguments involved. [O3]

Plenary suggestions

Write or give a talk Ask the students to write a paragraph or prepare a talk, starting with the following sentence: 'I would not like to live near a mine because...'. [O1, 2&3]

Answer questions Select questions from the sticky notes written at the start of the lesson and see how many can be answered by students in a given time. [O1, 2&3]

Answers to Student Book questions

1. they can remove waste rock and plant the area with trees and bushes, to hide the impacts of scarring
2. mines cause subsidence; their houses may be at risk of collapse
3. heavy machinery and trucks burn fossil fuels, which release carbon dioxide into the atmosphere; mines need electricity, which indirectly uses fossil fuels and releases carbon dioxide; carbonate ores need to be heated and decomposed, releasing carbon dioxide
4. either reduction of noise; or release of gases like carbon dioxide; it is hard to contain the release of gases particularly from moving vehicles, so they escape into the atmosphere; noise is carried by air particles; large thick barriers would have to be installed and these would be very costly
5. *Advantages*: no need to mine new materials, so conserving supplies of metal ores; no need to mine new materials, so no more noise from blasting and traffic; less pollution through production of greenhouse gases and acid rain; no acid mine leachate, or hazardous waste; less energy is used to recycle metals than to produce new ones.

   Disadvantages: recycling needs to be planned, so that waste is organised in the home; and the metals can be dealt with separately from other waste; recycling processes use lots of energy; and can result in toxic waste products; not everyone likes to make the effort to recycle.
6. For example: have separate bins for recycling different metals; improve public understanding of the importance of recycling and the need to keep metals separate; only recycle metals where is a definite environmental benefit, excluding other metals because of the energy needed to separate them.

Answers to Worksheet 3.3.7

1. a), b) for example:

   i) *Impacts on the land*: problems of waste rock; aesthetic impacts; subsidence; large hole to fill when the mine is no longer productive.

   ii) *Impacts on the air*: CO_2 released from the burning of fossil fuels to run heavy trucks; CO_2 released from the running of machinery; CO_2 released in using carbon to extract metals from their ores; sulfur dioxide released when ores containing sulfides are roasted; dust formed during the blasting process; lots of noise produced.

   iii) *Impacts on the water*: leachates of toxic materials from the rock, including heavy metals that could enter rivers and lakes; leachates from the mining process can enter rivers and lakes.

   iv) *Solutions* include: reclamation by planting trees and plants; specially lined pools to store leachates; water sprays to reduce dust; recycling metals
2. a) the second paragraph of the text is about the problems with recycling; the remainder is about the benefits.

   b) the main themes around the consequence map should include the mixture of benefits and problems set out in the text
3. Students should show an awareness that if there were no metals, it would be hard to conduct electricity and heat, so any appliances depending on these would not be in use; that metals are very strong and, although materials are being developed to replace them, like carbon fibre, these are very expensive, so a world without metals would have fewer bridges, trains, planes and cars.

3.8 Applying key ideas

Protecting metals

Objectives

- Extract ideas relating to how metals are used to protect each other.
- Apply ideas about the reactivity series, writing word equations for displacement reactions.
- Use ideas and information about particles to write balanced symbol equations and write detailed explanations of metal extraction processes.

Outcome

- Clear and effective responses to questions, indicating understanding and the next steps in learning.

The purpose of this activity is to provide opportunities to see how successfully students are grasping the key ideas so far. It isn't designed to be used as a formal test – it might be that students work on the questions collectively. It does provide an opportunity for the teacher to look at written work, engage students in discussions and form ideas about progress being made.

The tasks are progressive. Lower-attaining students should be able to tackle the first two or three tasks and higher-attaining students will find the later ones more challenging.

Resources needed word cards from Worksheet 3.3.4 to make up reactions

Teaching and learning

Engage

- Provide the students with the word cards about different elements and compounds from Worksheet 3.3.4. Ask them to make up three of their own different displacement reactions.

  Higher-attaining students should be able to **write balanced symbol equations** for these.
- Ask students to talk in pairs for a minute about everything they know about metal extraction.

Challenge and develop

- Show videos to demonstrate the process of anodising. If possible, show samples of anodised steel and un-anodised steel. Read paragraphs 2 and 3 in the 'Metal protection' section in the Student Book relating to anodising. Discuss the process, how it works and some of its uses. Ask the students to **compare and contrast** this with the thermit reaction. Ask them to **explain** how both processes use the reactivity series in different ways.
- Show a video to demonstrate the process of galvanising. Ask the students to **compare** this with the process of anodising.
- Finally show a demonstration involving dipping copper coins in silver nitrate solution. Leave them there for a couple of minutes and a silver coating will be deposited on the surface of the copper. Discuss the reaction.

Explain

Ask the students to read the rest of the 'Metal protection' section in the Student Book. Depending on their ability to assimilate text you may need to adopt strategies such as:

- Use the text as a source of ideas, draw a spider chart with 'Metal protection' at the centre and the different methods and applications grouped round it.
- Draw up a table to compare and contrast the different methods of protecting metals.
- Work with any groups of students that have difficulty with reading to support them in decoding the text and accessing the ideas.

Consolidate and apply

- Now ask the students to attempt the tasks – they might do this individually or collaboratively. In each case, encourage them to **identify** and **record** their ideas.
- Ask them to **present** their responses to various tasks, either orally or by displaying their work.

Extend

- For students able to progress further, task 5 gives an opportunity for extension work.

Plenary

Ask the students the following questions to reflect on how well they responded to the tasks:

- What did you find easy to do?
- What was hard?
- What do you still not understand about the reactivity series, word equations, balanced symbol equations and metal extraction?

Provide the class with some overall feedback, indicating ideas that they have grasped effectively and those that may need developing further.

Likely responses and next steps in learning

1. This task gives the students an opportunity to engage with the text and to extract some definitions from it. They should appreciate that anodising involves coating any metal with aluminium; galvanising involves coating less reactive metals with zinc and that sacrifical protection is shown. There is some sacrificial protection with aluminium, although the formation of the oxide layer is the main form of defence because this is resistant to further corrosion.
2. This task gives an opportunity for students to read about a situation and extract the chemical meaning from it. By deriving word equations, students are summarising the changes that occur. They should show the formation of aluminium oxide in the word equation and the displacement of silver to make silver and copper nitrate.
3. In this task the students have to apply information about anodising to this new context. They should recognise that once an aluminium layer has been plated on the magnesium it will react with air to make an oxide layer. This oxide layer provides sufficient resistance to further attack because it is very durable and strong. The students should recognise that anodising can be used with metals that are either more or less reactive than aluminium.
4. In this task the students need to demonstrate their ability to read each situation, recognise what changes are occurring and write the correct balanced symbol equations. In the first change, zinc is sacrificed and reacts with oxygen in place of the iron. In the second, aluminium displaces copper from the copper nitrate solution. This is a hard reaction to balance – they may need some support in realising that the nitrate group should be counted as one entity. In the third reaction, tin reacts with oxygen in place of the steel. The equations are:

   $2Zn + O_2 \rightarrow 2ZnO$

   $2Al + 3Cu(NO_3)_2 \rightarrow 3Cu + 2Al(NO_3)_3$

   $Sn + O_2 \rightarrow SnO_2$
5. The students should present a clear and logically ordered explanation of how carbon is used to extract each of these metals in turn from their ores in this final task. They should appreciate that zinc ore may be in the form of the sulfide and show how this needs to be roasted to make the oxide. Some detail of froth flotation is expected. They are also expected to explain how these ores vary in their ease of extraction and link this to their reactivity.

   They should be aware that aluminium is more reactive than carbon and cannot be extracted in the same way. Balanced symbol equations are expected for each step in the extractions.

3.9 Understanding exothermic reactions

Lesson overview

Learning objectives

- Describe examples of exothermic reactions.
- Explain the energy changes taking place during an exothermic reaction.

Learning outcomes

- Describe exothermic reactions, with examples from metal extraction including the thermit reaction. [O1]
- Explain the energy changes taking place during an exothermic reaction. [O2]
- Use energy-level diagrams to compare the energy in the reactants and products of an exothermic reaction and explain the energy changes in the particles. [O3]

Skills development

- Thinking scientifically: ask questions
- Working scientifically: record observations
- Learner development: collaborate effectively

Resources needed sticky notes; equipment and materials as detailed in the Technician's notes; Worksheet 3.3.9; Practical sheet 3.3.9; Technician's notes 3.3.9

Digital resources Quick starter; Interactive activity: Reorder the sentences to describe what happens when substances react

Common misconceptions All reactions give out heat.

Key vocabulary exothermic, bond-making, bond-breaking

Teaching and learning

Engage

- Demonstrate the reaction between potassium permanganate and glycerol, following the procedure and safety notes in Technician's notes 3.3.9. Safety screens are needed. Alternatively show a video of the reaction. Ask the students to observe and then **devise** some questions about the reaction on sticky notes. They should **describe** what they see happening. [O1]
- Ask the students to **recall** other reactions in which they can **identify** that energy is given out. Make up a class list. [O1]
- The students should **devise** further questions about exothermic reactions on sticky notes from their observations. [O1]

Challenge and develop

- **Group work** Arrange the students into small groups. Use the 'Examples of exothermic reactions' section in the Student Book to introduce the term 'exothermic'. Ask the students to follow the instructions on Practical sheet 3.3.9 and make detailed **observations** of the reactions – they should **determine** how exothermic the different reactions are. [O1]
- Select a few students from different groups to **present** their results. [O1]
- Use **role-play**, 'molymods' and the Student Book to explain and explore the progress of a chemical reaction, including how the reactant molecules collide and take in energy from the collision and their surroundings to break reactant bonds. Discuss what happens when new bonds are made – energy is released to the surroundings. Establish that a reaction is exothermic if the energy given out during bond-making is greater than the energy taken in during bond-breaking. [O1&2]
- The students should answer questions 1–4 from the Student Book. [O1&2]

Higher-attaining students should be able to **interpret** energy diagrams as shown in the Student Book to **explain** how an exothermic reaction occurs. They should answer questions from all three sections of the Student Book. [O3]

Explain

- **Group work** Ask the students, in their groups, to select two of the reactions they have undertaken. Ask them to use their own models or role-play to **explain** why one reaction is more or less exothermic than another. They should include ideas about bond-breaking and bond-making. [O1&2]

Consolidate and apply

- **Pair work** Arrange the students in pairs. Show a video demonstration of the thermit reaction, and link this to the idea of displacement they are already familiar with.

  Ask middle- and lower-attaining students to **write a word equation** for the reaction and **identify** the type of reaction occurring. [O1&2]

  Higher-attaining students should **write a balanced symbol equation**.

- The students should complete the tasks of Worksheet 3.3.9. [O1, 2&3]

Extend

- Ask students able to progress further to **sketch** energy-level diagrams for all the reactions demonstrated and carried out. Their diagrams should reflect the degree to which the reactions are more or less exothermic. [O3]

Plenary suggestions

Design a poster Ask the students, in small groups, to **design** their own heat pack on a poster. They should carefully **select** one of the reactions they have used in the lesson and **justify** why they have chosen it. They should be able to **explain** why the reaction is exothermic. [O1, 2&3]

Questions and answers Choose different questions, posed at the start of the lesson, from some students and select other students to answer them. [O1, 2&3]

Answers to Student Book questions

1. thermit reaction; it has the highest energy change
2. an explosion; a firework – particularly a 'banger'
3. the magnesium and oxygen molecules collide; the bonds between the magnesium atoms and the bonds between the oxygen atoms use energy from the collision and the surroundings to break; the atoms are now free to make new bonds; as they make magnesium oxide, energy is released to the surroundings; because more energy is released during bond-making compared to the energy taken in during bond-breaking, the reaction is exothermic
4. The same amount of energy is taken in during bond-breaking as is released during bond-making.
5. The sketches should show that the gap between the energy of the reactants and the energy of the products is vastly higher in the thermit reaction, compared to the neutralisation reaction.

Answers to Worksheet 3.3.9

1. a) B, C, D, A b) C c) B d) D
2. The drawings should include a collision between the iron oxide and aluminium. The bonds between the iron and oxygen and aluminium and aluminium should be broken; and the atoms should be shown to be free. The students should recognise that aluminium is a metal; and that metallic bonding occurs. Energy should be shown to be taken in during this process. New bonds will form between the aluminium and oxygen; leaving the iron free; this process should be shown to give out much more energy than the energy taken in.
3. the biggest difference between the energy of the reactants and products should be shown for reaction B; there is no difference in energy between the reactants and products for reaction A; a small difference in energy difference should be seen between the reactants and products for reaction D; a slightly larger one occurs for C

3.10 Comparing endothermic and exothermic changes

Lesson overview

Learning objectives

- Describe examples of endothermic reactions.
- Compare the energy changes during exothermic and endothermic reactions.

Learning outcomes

- Describe endothermic reactions, with examples; identify factors to control, change and measure in the investigation of an endothermic reaction. [O1]
- Draw energy-level diagrams to represent endothermic reactions and exothermic reactions; explain how to make precise, accurate and reliable measurements. [O2]
- Use energy-level diagrams to explain changes taking place during an endothermic reaction and compare these with an exothermic reaction. [O3]

Skills development

- Thinking scientifically: consider the quality of evidence
- Working scientifically: design investigations
- Learner development: collaborate effectively

Resources needed a green plant; a cold pack (containing ammonium nitrate); equipment and materials as detailed in the Technician's notes; Worksheet 3.3.10; Practical sheet 3.3.10; Technician's notes 3.3.10

Digital resources Quick starter; Interactive activity: Drag the statements to the correct group - exothermic or endothermic change

Common misconceptions The temperature does not fall during an endothermic change.

Key vocabulary endothermic

Teaching and learning

Engage

- Show the students a plant photosynthesing and a cold pack. Ask them to **identify** what these two changes could possibly have in common. Introduce the idea of an endothermic reaction as the opposite of an exothermic reaction. Ask the students to **predict** what would happen to the temperature during an endothermic reaction. [O1]

Challenge and develop

- **Group work** Arrange the students into small groups. Ask them to **plan an investigation** to find out if the amount of the reactant affects how far the temperature falls in the reaction between citric acid and sodium hydrogencarbonate. They should answer the planning questions given on Practical sheet 3.3.10. [O1&2]
- Select different groups to **share** their ideas on how to make the evidence accurate, precise and reliable. [O2]
- When their plans have been checked, ask the students to **carry out** their investigations and **collect evidence**. They should **draw a graph** of their data, and answer the analysis question on the Practical sheet. [O1, 2&3]

Explain

- Select different groups to **explain** their findings.

  Lower-attaining students should **explain** their choice of variables. [O1]

  Middle-attaining students should **draw energy-level diagrams** to describe the difference between each of their results. [O2]

Higher-attaining students should use ideas about bond-breaking and bond-making to **explain** their results. [O3]

Consolidate and apply

- Show a video of the highly endothermic reaction between barium hydroxide octahydrate and ammonium chloride. Discuss the large temperature change in the reaction. [O1&2]
- Students should **describe** the differences between the reaction they carried out themselves and the reaction observed between barium hydroxide and ammonium chloride. [O1&2]

  Higher-attaining students should **compare and contrast** the reaction between barium hydroxide and ammonium chloride with an exothermic reaction from last lesson, using energy-level diagrams. [O3]
- Students should work through the appropriate questions in the Student Book and the and tasks of Worksheet 3.3.10. [O1, 2&3]

Extend

- Ask students able to progress further to use the results from their investigations to **produce a detailed plan** of an investigation into how to make the best cold pack using ammonium nitrate or citric acid and sodium hydrogencarbonate. They should use ideas about bond-breaking and bond-making to **justify** the choices they make for the values of their independent and control variables. [O3]

Plenary suggestions

True or false? Statements like these could be given as a true/false quiz: [O1, 2&3]

- Exothermic reactions are used in cold packs (F)
- More energy is used to break bonds in endothermic reactions (T)
- Photosynthesis is the most common endothermic reaction (T)
- Heat packs can use citric acid and sodium hydrogencarbonate (F)
- More reactions are exothermic compared to endothermic (T)

Answers to Student Book questions

1. Endothermic reactions take more energy in than they give out.
2. slower; there is less energy to absorb from the surroundings
3.

Exothermic reaction	Endothermic reaction
More energy is released when bonds are made	More energy is absorbed when bonds are broken
Temperature increases	Temperature decreases
Used for heat packs	Used for cold packs
Energy can be released as heat, light or sound	Energy is mostly absorbed as heat
Burning magnesium or other suitable example	Ammonium nitrate and water or other suitable example

4. The sketch should show the energy of the products being higher than the energy of the reactants in both graphs; for the more endothermic reaction (barium hydroxide and ammonium chloride) there should be a larger difference.
5. a) exothermic b) exothermic c) exothermic d) endothermic
6. Some elements make stronger bonds than others.

Answers to Worksheet 3.3.10

1. answer should include: change the amount of ammonium nitrate (independent variable); measure the temperature change (dependent variable); keep the amount of water the same (control variable); the amount to induce a temperature change of between 10 and 20 °C would be suitable
2. the energy of the products should be shown higher than the energy of the reactants for both reactions; for the much more endothermic reaction (barium hydroxide and ammonium chloride) there should be a much larger difference; suitable annotations
3. Students should explain that much more energy is taken in for the bond-breaking of barium hydroxide and ammonium chloride reaction than is released by bond-making; there is a smaller difference between bond-breaking and bond-making for the reaction between ammonium nitrate and water.

3.11 Explaining the use of catalysts

Lesson overview

Learning objectives

- Describe what a catalyst is.
- Explain how catalysts work.

Learning outcomes

- Describe what a catalyst does; give examples of different catalysts. [O1]
- Interpret data to explain how a catalyst affects a reaction. [O2]
- Explain how a catalyst works. [O3]

Skills development

- Thinking scientifically: consider the quality of evidence
- Working scientifically: design investigations
- Learner development: collaborate effectively

Resources needed a sealed test tube of 10 vol hydrogen peroxide that has been left overnight; equipment and materials as detailed in the Technician's notes; Worksheet 3.3.11; Practical sheet 3.3.11; Technician's notes 3.3.11

Digital resources Quick starter; Slideshow: Catalysis of hydrogen peroxide decomposition - A method to compare catalytic effectiveness; Interactive activity: Match the catalyst to its use

Common misconceptions Only heat can make a reaction go faster.

Key vocabulary catalyst, rate of reaction, enzyme, catalytic converter

Teaching and learning

Engage

- Show the students a sealed test tube of hydrogen peroxide that has been left overnight. Explain that it decomposes very slowly to make water and oxygen. Test the gas in the test tube with a blown-out splint for oxygen – there should be no sign of it. Now add a spatula of manganese dioxide to a conical flask containing hydrogen peroxide and ask the students to **observe** the changes. Ask some students to feel the flask and test the gas for the presence of oxygen. If you add a few drops of washing-up liquid to the hydrogen peroxide, as detailed in the procedure in Technician's notes 3.3.11, the production of oxygen will cause a foam to be made. Ask the students what has happened to the manganese dioxide at the end of the reaction. Filter off the manganese dioxide and use it in another fresh flask of hydrogen peroxide. The students should realise that the manganese dioxide works just as well – it has not been used up. [O1]

Challenge and develop

- Use the Student Book to introduce, discuss and define the term 'catalyst'. Discuss different examples of catalysts and how they affect chemical reactions. Discuss enzymes as biological catalysts and their importance to life. [O1]

- **Group work** Arrange the students into small groups. Tell them that an enzyme called catalase also catalyses the decomposition of hydrogen peroxide. It can be found in potatoes, apples and liver. Ask the groups to **plan an investigation** to see which of the three contains the most catalase and catalyses the reaction most effectively. They can use Practical sheet 3.3.11 to guide them. Demonstrate how to set up the equipment first and discuss how to measure the rate of the reaction. The reactants can be put in a large measuring cylinder with a few drops of washing-up liquid. A foam will be produced, and the height of this can be measured in the measuring cylinder. The measuring cylinder should stand in a tray to avoid spillages. The students should **consider** how to collect accurate and reliable data. [O1&2]

- When their plans have been checked, the students **carry out** their investigation, **make observations** and **record results**. [O1&2]

Explain

- Ask all the students to **draw graphs** of their data and to **explain** their findings. They should **conclude** which is the most effective catalyst and why. They should **evaluate** their evidence, answering all the analysis and evaluation questions on Practical sheet 3.3.11, **explaining** how their investigation could be improved. [O1&2]

  Higher-attaining students should include ideas about how the catalyst works. [O3]

Consolidate and apply

- The students should answer all the questions in the Student Book and complete the tasks of Worksheet 3.3.11. [O1, 2&3]
- They could **explain** the similarities and differences between manganese dioxide and catalase. [O1&3]

Extend

- Ask students able to progress further to **calculate** the rate of the reaction from the graphs they have drawn and **plot** the rate against time for the different catalysts. They could then use these graphs to **compare** the catalysts in a quantitative way. [O3]

Plenary suggestions

Mind maps Ask the students to create their own mind maps to describe what catalysts are, explain how they work and describe common uses. Key questions for the mind map could include 'What are catalysts?', 'What do catalysts do?', 'How do catalysts work?'. [O1, 2&3]

Answers to Student Book questions

1. try to recover it after the reaction; see if it has the same effect with fresh reactants
2. so that useful products can be made in a shorter time
3. curve A; more oxygen gas is made in a certain time compared to curve B; graph B indicates no reaction
4. carry out a reaction with the same mass of manganese dioxide and catalase; put them in identical conical flasks, with 25 cm^3 of hydrogen peroxide in each. Measure the amount of oxygen produced in a given time, and compare the results
5. they should show an enzyme molecule with a specific shape, locking onto the substrate and keeping it in place for water molecules to break down the substrate
6. Catalysts should be used in powdered form, because they have a higher surface area, so the reaction will proceed even faster. In a large lump the catalyst will have a smaller surface area, and the reaction will not proceed as quickly.

Answers to Worksheet 3.3.11

1. b; enzymes are biological catalysts; catalysts can be used again, because they do not get used up in the reaction
2. a) two graphs should be drawn; on the same axes; the time should be on the *x*-axis, and the volume of gas on the *y*-axis; appropriate scales should be used for both axes

   b) the graph drawn using data set 2 is the catalysed reaction; more gas is produced per unit of time
3. The diagram should show the platinium/rhodium plate with a large surface area, holding the NO molecules and enabling them to be broken down. When the oxygen atoms are released they react with CO on the catalyst to make CO_2 and with methane (CH_4) on the catalyst to make CO_2 and H_2O; the nitrogen atoms join together to make nitrogen gas (N_2).

3.12 Exploring ceramics and their properties

Lesson overview

Learning objectives

- Describe what is meant by the term ceramic.
- Describe the properties of ceramics.

Learning outcomes

- Describe some examples and uses of common ceramics. [O1]
- Explain how different types of ceramics vary in their properties. [O2]
- Explain how the chemistry and bonding in a ceramic affects its properties. [O3]

Skills development

- Thinking scientifically: ask questions
- Working scientifically: design investigations
- Learner development: collaborate effectively

Resources needed a range of objects made from different materials including plastics, metals, ceramics and composites; selection of different ceramics including tiles, bricks, plates, glassware, vases; Worksheet 3.3.12; Practical sheet 3.3.12

Digital resources Quick starter; Interactive activity: Drag the statements to the correct group - ceramic, plastic or metal; Video

Common misconceptions Clay pots are the only type of ceramic.

Key vocabulary ceramic, crystalline, amorphous

Teaching and learning

Engage

- **Pair work** Show the students a range of objects made from different materials including plastics, metals, ceramics and composites. Ask them to **discuss** in pairs what each object is made from and to **categorise** the objects in terms of the materials they are made from. Leave these objects in place until the end of the lesson. [O1]
- Show a video of different types of ceramics, or use the Student Book. Discuss what is meant by the term 'ceramic' and what the two main types are. [O1]

Challenge and develop

- **Group talk** Working in small groups, the students **discuss** how clay is made and **explore** the different independent variables that could be investigated to change the properties of clay – these could include the ratio of clay powder to water, the temperature at which the clay is baked, the length of time for firing the clay, the types of clay used. [O1, 2&3]
- Ask the groups to **plan an investigation** to make some clay and to investigate how the ratio of clay powder to water affects the properties of the final ceramic. Discuss the dependent variable – how will they measure the effect of different compositions on the properties of clay? The students can use Practical sheet 3.3.12 to do this, answering all the questions to help them with their plan.

  Clay is difficult and messy to use – and if a kiln is not available then it is not realistic to carry out this investigation. However, the students will benefit from thinking about how to carry out the investigation, developing a hypothesis and how to test it. [O1, 2&3]

- Now ask the groups to **devise** their own survey of different ceramics to **identify** their properties – they should use the 'General properties of ceramics' section in the Student Book to help them design the survey and use a scoring system so they can **compare** the properties of one ceramic with another. [O2&3]

  Higher-attaining students should **identify differences** between crystalline ceramics and amorphous ceramics.

Explain

- Ask different groups to **explain** their findings about ceramics and their properties to the class. They should explain the system they used for comparing the properties of different ceramics. [O2]
- The students should answer Student Book questions 1–4. [O1&2]

  Higher-attaining students should **explain** what they have found out about the differences in properties between amorphous and crystalline ceramics. They should answer questions 5 and 6 in the Student Book. [O3]

Consolidate and apply

- Ask the students to complete the tasks of Worksheet 3.3.12. [O1, 2&3]
- **Pair work** Ask pairs of students to **imagine** they have found a new material. Ask them to **devise a flowchart** to identify if it is a ceramic or not. [O1&2]

Extend

- Ask students able to progress further to carry out further **research** into how the composition and chemistry of ceramics affects their properties. [O3]

Plenary suggestions

Make a poster The students could create a small poster to define what a ceramic is and the factors that affect its properties. [O1, 2&3]

Grouping objects Return to the objects used at the start of the lesson. Ask the students to change any categorisation of the objects and to give one property of the ceramics they identify, including their structure. [O1, 2&3]

Answers to Student Book questions

1. these could include toilets; plates; bricks; tiles; mirrors etc.
2. they are very long lasting; they do not decay easily
3. *ceramics*: hard; light; resistant to wear and corrosion; brittle; thermal and electrical insulators; non-magnetic; non-toxic; not ductile

   metals: hard; mostly heavy; mostly corrode; not brittle; hard to break; thermal and electrical conductors; some are magnetic; some are toxic; most are ductile
4. they have many properties in common, but ceramics are cheaper to make than metals
5. *crystalline*: three-dimensional pattern in the bonding of atoms; heated to high temperature; permanent structure

   amorphous: no regular crystalline shape; heated then cooled quickly; shape can change very slowly over long periods of time
6. there is more variability in the composition of the ingredients from nature, for example clay has different chemicals depending on where it comes from; this makes the properties less predictable and uniform

Answers to Worksheet 3.3.12

1. b, d, e, f
2. *tiles* – brittle; resistant to wear; chemically stable; quite hard to scratch; quite dense

   glass – very brittle; easily scratched; quite light

   porcelain – resistant to wear; chemically stable; very hard to scratch; quite light

   brick – easy to scratch; not very brittle; resistant to wear; quite dense
3. The diagram should show that glass has no regular structure, therefore its particles are less strongly held together, making it easier to scratch and more brittle.

3.13 Matching properties of ceramics to their uses

Lesson overview

Learning objectives

- Explain how the properties of ceramics determine their uses.

Learning outcomes

- Describe some main uses of different types of ceramics; present data about the properties of ceramics. [O1]
- Interpret data to find patterns relating to the properties of ceramics. [O2]
- Explain the advantages and disadvantages of using ceramic materials. [O3]

Skills development

- Thinking scientifically: analyse data
- Working scientifically: present data
- Learner development: communicate effectively

Resources needed sticky notes; graph paper; pieces of glass, tile, porcelain and brick; Worksheet 3.3.13

Digital resources Quick starter; Slideshow: Ceramic materials: What are they, and where are they used?; Interactive activity: Place the materials, in order, from highest to lowest melting point

Common misconceptions Ceramics are only used as household objects.

Key vocabulary earthenware, stoneware, porcelain, refractory ceramics

Teaching and learning

Engage

- Show the students a range of different ceramics. Tell them that ceramics can be classified according to their uses. Ask them to **classify** the objects accordingly. [O1]
- Use the first section in the Student Book to explore the range of different uses of ceramics. [O1]

Challenge and develop

- **Group talk** Arrange the students into their groups from the previous lesson. Give each group pieces of glass, tile, porcelain and brick. Ask the students to apply their survey from the previous lesson to the samples. Question them about how well the survey and scoring system they devised worked for their samples. [O1&3]
- Ask the groups to **discuss** ways of presenting their own data and that from the Student Book, as suggested in task 1 of Worksheet 3.3.13. [O1]
- Give students graph paper and ask them to **plot** the graphs suggested in the differentiated tasks of Worksheet 3.3.13. They can then **display** them on a poster. They should **annotate** the strengths and weakness of their graphs, and describe what the data is telling them. [O1, 2&3]

Explain

- Ask the students to **present** their posters and **explain** what their data is telling them. Encourage the class to **pose questions** to each group about their data, relating to the quality or to the findings. [O2]

Consolidate and apply

- Ask the students to **draw a concept map** of the different uses of ceramics. They should use their data and include this in the map. They should **peer-assess** each others' concept maps, using sticky notes, and identify any features that are missing. [O1&2]

Higher-attaining students should use the Student Book to **annotate** their concept maps, **explaining** the advantages and disadvantages of using ceramics. [O3]

- The students should complete the questions from the Student Book. [O1, 2&3]

Extend

- Ask students able to progress further to **compare** the properties and uses of early ceramics with modern-day ones. They should **identify** how the structure and bonding may have changed, and what modern-day ceramics are mainly used for compared to the earthenware used in earlier times. They should **evaluate** their advantages and disadvantages – they could **present** their ideas to the class. [O3]

Plenary suggestions

'Trump cards' The students could make some 'trump cards' about ceramics and their properties. [O1&2]

Write a summary They could write a summary of what ceramics are, what their properties are and what they are used for.

Higher-attaining students could include problems associated with the use of ceramics. [O1, 2&3]

Answers to Student Book questions

1. waterproof; non-toxic; hard; easy to make into different shapes
2. all uses could relate to other materials, but ceramics are very commonly the main type of material used
3. ceramics are harder than metals; they have higher melting points; they do not expand as much on heating
4. a) melting point; expansion on heating

   b) silicon carbide has the highest melting point and the lowest expansion on heating
5. they are mainly mixtures of different compounds – the materials are not chemically combined to form a product; they are not combined in an exact ratio; in compounds the atoms are combined in an exact ratio
6. ceramics are good electrical and thermal insulators whereas metals are conductors. Ceramics can compete with metals in many applications, but not those requiring electrical and thermal conduction

Answers to Worksheet 3.3.13

1. a) i) name of material against any of the following: melting point; hardness; density; measure of expansion on heating

   ii) any of: hardness v density; hardness v melting point; hardness v expansion on heating; and vice versa; melting point v density; melting point v expansion on heating; and vice versa; density v expansion on heating; and vice versa

   iii) none

   b) a bar graph should be drawn with the name of the material on the x-axis and the hardness on the y-axis, with a scale of 1–9 equally spaced; the bars should be equally spaced
2. appropriate bar graphs with the name of the material on the x-axes; the bars should be equally spaced and take up the same amount of width; the y-axis scales should be appropriate, making the most of the graph paper. The patterns shown should be clearly described – ceramics have higher melting points than metals; they are less dense than steel, but more dense than aluminium
3. a) i) melting point and density – there is no correlation; students may plot either as the independent variable; scales should be equally spaced and good use should be made of the graph paper

   ii) expansion on heating and melting point – negative correlation: as the measure of expansion increases, the melting point decreases; a line of best fit should be drawn, with anomalous points suitably highlighted

   b) i) Metals are often mined from deep within the Earth, and extracted from rocks using different chemicals, some of which are toxic. Traditional ceramics are made from clay, which is taken from near the surface of the Earth. There is little processing to be done to use it. More energy is needed and more greenhouse gases released in extracting metals compared to making traditional ceramics. Modern ceramics are compounds of metals so have all the disadvantages associated with metal extraction.

   ii) the bonding in ceramics is likely to be stronger; the high melting points and hardness data show this

3.14 Exploring natural polymers

Lesson overview

Learning objectives

- Explain what a polymer is.
- Describe examples of natural polymers.

Learning outcomes

- Describe what is meant by the term 'polymer'; design an investigation to determine the properties of different polymers. [O1]
- Describe the properties of natural polymers; collect accurate and reliable evidence about the properties of different natural polymers. [O2]
- Evaluate an investigation design and suggest improvements. [O3]

Skills development

- Thinking scientifically: ask questions
- Working scientifically: design investigations
- Learner development: collaborate effectively

Resources needed equipment and materials as detailed in the Technician's notes; Worksheet 3.3.14; Technician's notes 3.3.14

Digital resources Quick starter; Interactive activity: Match the monomer to the polymer; Slideshow: Some natural polymers

Common misconceptions Plastics are the only type of polymer.

Key vocabulary polymer, monomer, protein, cellulose, chitin

Teaching and learning

Engage

- Show the students samples of materials such as wood, rubber, silk, cotton, hemp, hair and starch. Ask them to **consider** what these materials have in common. Explore the ideas that many chemicals from living things are not made up of simple molecules and consist of very long chains called polymers. [O1]
- The students should read the 'What is a polymer?' section in the Student Book and **explain** the concept of a polymer. [O1]

Challenge and develop

- **Small groups** The students work in groups to **devise a model or role-play** to explain what a polymer is and how it is different from other chemicals they know of. [O1]
- Still in their groups, they should work together to **design an experiment** to investigate the properties of natural polymers such as wood, rubber, silk, cotton, hair and hemp. They should complete the **planning** task on a poster. They need to **consider** what kinds of properties they could investigate and how they will collect the evidence. They should **recall** the prompts from recent planning tasks and try to plan this investigation independently of any support. (No practical sheet is provided.) They could investigate the density, elasticity, solubility, hardness or strength of the polymers. [O1]
- When students have completed their draft plans, ask them to **identify** how they will make their investigations repeatable, reproducible, valid and accurate. They should add annotations to their designs using highlighter pens to **explain** these points. They should write any procedures clearly enough for others to follow. [O1]
- When you have approved their final plans, the students can **carry out** their procedure, **collect evidence** and **record** their results in clear tables. [O2]

- If there is time available, ask each group to swap procedures with another group and **follow their plan** to collect evidence and record results. [O2]
- The students should **evaluate** the new plans they have been given and **suggest improvements** that could be made. [O3]

Explain

- Select two groups who carried out the same procedures. They should, in turn, **present** and **compare** their results with the class and **discuss** how reproducible the procedures were. The class can **identify improvements** that could be made to each plan. [O1&3]

Consolidate and apply

- Ask the groups to **set targets** for themselves in terms of developing their skills in experimental design. They should identify what they need to do as individuals to **improve their skills**. [O1&3]
- The students should answer the questions from the Student Book and complete the tasks of Worksheet 3.3.14. [O1, 2&3]

Extend

- Ask students able to progress further to **research** other polymers that come from plants. They should find out what the starting monomer is for all chemicals in plants. [O3]

Plenary suggestions

True or false? Here are some suggested statements for a true/false quiz: [O1, 2&3]

- Polymers are found only in living things (F)
- Sugar is a type of polymer (F)
- Investigations should always be reproducible (T)
- To make an experiment reproducible, you need to repeat it yourself (F)
- If the control variables are not controlled the experiment is not valid (T)

Answers to Student Book questions

1. chemical
2. so that soluble monomers can be converted into insoluble polymers, and be safely stored inside cells
3. keratin
4. it is not soluble; it is made from the same elements as sugar
5. no; they have been made by chemical reactions involving the monomers; they have different structures and are not chemically the same
6. the model should show a polymer with long strands that are entangled, and can be elongated to form longer strands when stretched

Answers to Worksheet 3.3.14

1. big; long chains; monomer; heat; pressure; catalyst; DNA; protein; cellulose; starch; cotton
2. 1 with C; 2 with D; 3 with E; 4 with B; 5 with A
3. The student should use the same length of each material to be tested – e.g.1 m. She should fix one end of the material and use a newtonmeter to apply a force of say 10 N to the other end. She should measure the extension of the material and record it. She should repeat this at least three times with a new 1 m length of the same material, to see if her results are reliable. If they are close enough, she should work out the mean value and use this as her result. She should repeat this with the other materials to be investigated, ensuring that she follows exactly the same procedure. Finally she should see how reproducible her data is, by asking someone else to carry out the same investigation, and then comparing results.

3.15 Using human-made polymers

Lesson overview

Learning objectives

- Describe how human-made polymers are made in simple terms.
- Describe uses for human-made polymers.

Learning outcomes

- Name some human-made polymers and their uses. [O1]
- Explain how the properties of polymers match their uses. [O2]
- Explain how the properties of polymers are affected by their bonding using simple models. [O3]

Skills development

- Thinking scientifically: ask questions
- Working scientifically: develop explanations
- Learner development: communicate effectively

Resources needed sticky notes; large sheets of paper; coloured pens; Worksheet 3.3.15 (page 2 copied onto card)

Digital resources Quick starter; Video; Interactive activity: Reorder the sentences to describe how to make polypropene film; Interactive activity: Match the polymer to its use

Common misconceptions Polythene is the only type of plastic.

Key vocabulary synthetic, polythene, PVC, cross-link

Teaching and learning

Engage

- Ask the students what is meant by the term 'plastics' and to **identify** any they are familiar with. If possible, show a video about the development of plastics. Establish that polythene was the first human-made polymer, but that many others developed as a result. [O1]

Challenge and develop

- The students should read the 'Making synthetic polymers' section in the Student Book and answer questions 1–2. [O1&2]
- Ask the students to complete the first two tasks of Worksheet 3.3.15. They should **match** the cards relating the properties of polymers to their uses. [O1&2]
- **Group work** Arrange the students into small groups. Ask them to carry out some **research** and use the Student Book and information from the Worksheet task 2. They should **develop** a detailed concept map of human-made polymers, their properties and uses on a large sheet of paper; they should then **produce** a slideshow presentation. They should **plan** how they will present their ideas and **allocate roles** within the group to carry out particular tasks. [O1&2]

  Higher-attaining students should **find out** about the bonding in polymers and include ideas about this in their concept map. They should answer questions 5 and 6 from the Student Book. [O3]

Explain

- Ask each group to **present** their slideshow and show their concept maps, **explaining** the uses and properties of different polymers. [O1, 2&3]

Consolidate and apply

- Show the students a range of different human-made polymers . These could include low-density polythene, high-density polythene, polystyrene, polyester, acrylics, nylon, polypropylene, and polyurethane. Ask them to **identify** key properties of the polymers and **suggest** some uses for them. [O1&2]

  Higher-attaining students may **make predictions** about the type of bonding within the polymers. [O3]

- The students can attempt task 3 of Worksheet 3.3.15. [O3]

Extend

- Provide students able to progress further with some information about uses and properties of particular polymers. Ask them to **use models** to suggest what the chemical structure might look like within them. [O3]

Plenary suggestions

Devise questions Ask the students to devise some questions on sticky notes for another group to answer (the questions should relate to the presentations they performed earlier). They should swap questions and answer those from another group. [O1, 2&3]

Answers to Student Book questions

1. they both consist of long chains of linked monomers
2. they had to first understand what they were made from; this required an understanding of atoms along with structure and bonding; that was developed only in the early 1900s
3. they are extracted from crude oil and are mostly comprised of carbon atoms and hydrogen atoms (hydrocarbons)
4. aircraft canopies; containers; water pipes
5. The model should include strong intermolecular bonds, to account for a high melting point, but chains of molecules that are spaced out to account for a low density.
6. Cross-linked polymers are like a lattice, inside which small medicine molecules can become trapped; the medicines are carried inside the polymer molecules to make their delivery more effective.

Answers to Worksheet 3.3.15

1. a) polystyrene b) polypropylene c) polyacrylamide d) polychloroethene
2. b) A polymer can have many diverse applications, relying on different properties. This is due to the fact that they can be 'designed' to have particular properties by altering the bonding associated with them.

   c) presentations should link applications of polymers with properties; for example, polypropylene is used to make film for food packaging because it is transparent, able to stretch and non-toxic
3. diagrams should show lots of branches in the low-density polythene, and lots of straight chains in the high-density polythene, with the chains closer together to account for the higher density

3.16 Explaining composites

Lesson overview

Learning objectives

- Explain what is meant by the term 'composite'.
- Describe some uses of natural composites.

Learning outcomes

- Describe what is meant by a composite. [O1]
- Describe the properties of natural composites. [O2]
- Use models to explain how a composite is constructed; use models to explain how the structure affects the properties. [O3]

Skills development

- Thinking scientifically: use nomenclature
- Working scientifically: make predictions
- Learner development: communicate effectively

Resources needed a brick; a concrete block; examples of natural composites like bone and wood; equipment and materials as detailed in the Technician's notes; Worksheet 3.3.16; Practical sheet 3.3.16; Technician's notes 3.3.16

Digital resources Quick starter; Slideshow: Composites - Examples of use; Interactive activity: Drag the statements to the correct group - human-made composites, natural composites, binder or reinforcer

Common misconceptions Composite materials are new types of material.

Key vocabulary composite, concrete, matrix, reinforcement

Teaching and learning

Engage

- **Pair talk** Show the students a modern house-building brick and a picture of ancient brick (such as Figure 3.3.16a in the Student Book). Ask them, in pairs, to **compare and contrast** the two objects. Tell them that ancient bricks were made of mud and straw – they are examples of early composite materials. Define 'composite' using the 'What are composites?' section in the Student Book. Ask the students to **consider** if the modern brick is a composite and what evidence they could gather to **support their point of view**. [O1]

Challenge and develop

- Show videos, real examples and/or use the Student Book to explore other examples of composites. [O1]
- **Group work** Arrange the students into small groups. They will **investigate** making different samples of concrete using different compositions of materials. They should follow the instructions on Practical sheet 3.3.16. They will **make up** different samples of concrete, and in the next lesson will test their strength. [O1&2]

Explain

- Ask the students to **make predictions** to **explain** how they think the different compositions will affect the properties of the concrete. [O2]

  Higher-attaining students should **make a model** of their different concrete samples to **explain** how they differ from each other – they should use these models to **explain** how the different samples might compare in their strengths. [O3]

Consolidate and apply

- The students should complete tasks 1 and 2 of Worksheet 3.3.16. [O1&2]

Higher-attaining students should complete task 3 of Worksheet 3.3.16. They should **choose** one composite from the Student Book and **develop** their own pictures/model to **explain** its structure. [O3]

- The students should answer the Student Book questions. [O1, 2&3]

Extend

- Ask students able to progress further to **compare** the structures of bone, concrete and wood. They should **develop models** to explain how the structure of each of these affects their properties. They should **predict** which is the strongest, and which has the highest density. [O3]

Plenary suggestions

Making sentences Show the students three words and ask them to make up a sentence incorporating all three. Some examples are:

- composite, lignin, cellulose
- concrete, sand, water
- bone, concrete, wood. [O1, 2&3]

Differences The students should explain how composites are different to polymers and ceramics. [O1, 2&3]

Answers to Student Book questions

1. a composite is made of two or more materials, combined to make a stronger material
2. diagram of mud on its own and diagram of mud with straw in it; the straw should be interspersed throughout the brick making a stronger material
3. a) Concretes are mixtures, so the composition of each sample will not be truly uniform.

   b) A large sample size of concrete samples should be investigated.
4. In *wood*, the binder is cellulose and the reinforcement is lignin; in *bone* the binder is collagen and the reinforcement is calcium phosphate.
5. The model should show that the Brazilian Cherry has much more lignin compared to the Douglas Fir.
6. the reinforcement – calcium phosphate

Answers to Worksheet 3.3.16

1. Ancient Greeks made the first early **composites**. They mixed straw with mud to make bricks. These bricks were **stronger** and **lasted longer** than bricks without straw. Concrete is a **composite**. It is made with limestone, clay, **sand** and gravel. Wood and **bone** are two composites found in nature.
2. a) the ratio of the mix of gravel, cement, sand and water used to make it

   b) Compressive strength is the force (per unit cross-section) needed to squash the concrete; tensile strength is the force (per unit cross-section) needed to pull it apart.

   c) Too much water makes a weaker concrete; too little water gives a concrete that cracks easily.
3. a) The models/diagrams should clearly identify the matrix and the reinforcement. The reinforcement should be randomly dispersed within the matrix, and the two materials should be clearly identifiable.

   b) Usually a high ratio of reinforcement to matrix increases the strength, but if there is not enough matrix material the composite can crack and become brittle.

3.17 Using human-made composites

Lesson overview

Learning objectives

- Explain how human-made composites were developed.
- Describe the properties and uses of human-made composites.

Learning outcomes

- Recognise some human-made composites. [O1]
- Describe some uses of human-made composites. [O2]
- Use models to explain how human-made composites are constructed; use data to compare composite materials. [O3]

Skills development

- Thinking scientifically: ask questions
- Working scientifically: record evidence
- Learner development: collaborate effectively

Resources needed samples of human-made composites, e.g. fibreglass, reinforced concrete block; samples of concrete made in the previous lesson; equipment and materials as detailed in the Technician's notes; Worksheet 3.3.17, Practical sheet 3.3.17; Technician's notes 3.3.17

Digital resources Quick starter; Interactive activity: Place the materials, in order, from least to most dense; Interactive activity: Matching metals and composites to their tensile strength mass ratio; Hangman: Key vocabulary game

Common misconceptions Composites only consist of particulate, not fibrous, materials.

Key vocabulary fibreglass, carbon fibre, cermet

Teaching and learning

Engage

- Use the Student Book, videos or real samples to show the students a range of human-made composite materials. [O1]
- Introduce the idea that human-made composites can be superior in some respects to metals and other materials. Ask the students to develop a class list of properties of human-made composites that make them superior to other materials. [O1]

Challenge and develop

- Show the students a sample of reinforced concrete and a concrete sample made in the last lesson, or use the images on Worksheet 3.3.17. Discuss the differences and ask the students to complete task 1 of Worksheet 3.3.17. Demonstrate how to measure the strength of a concrete sample as shown on Practical sheet 3.3.17. [O1]
- **Group work** Ask the students to **follow the procedure** of the demonstration and **test** the strengths of all the concrete samples they made in the previous lesson – then they can **share** their findings. Ask them to use the data on the cards from Worksheet 3.3.17 and **compare** the strengths of their concrete samples with the data for reinforced concrete. [O1&3]
- **Expert groups** The students should be in groups of about six. Each group will become expert on a particular type of human-made composite, as outlined in task 2 of Worksheet 3.3.17. Groups could be given different colours to represent each of the composites they are learning about.

Reinforced concrete or cermet might be most suitable for lower-attaining students. Fibreglass might be most suitable for higher-attaining students.

Ask the groups to use the information cards from Worksheet 3.3.17 to **produce a spider diagram** about their composite. They should then use all the information they collate to **write a summary** about their composite. [O1, 2&3]

Explain

- **Rainbow groups** Ask one person from each group to form a new group, so that each group colour is represented in this new group. Ask the students to **teach** each other about the composites they have learned about. They should **prepare** a group poster about human-made composites, but ensure that they **present information** about a composite they have learned about from someone else. [O1, 2&3]

Higher-attaining students should **include a model** or annotated diagram of the composites in their part of the poster. [O3]

Consolidate and apply

- Ask each group to **present** their posters. The class should **listen** to the presentation and **devise** some questions they would like to ask. After the presentation, the group presenting should answer the questions they are posed. [O2]
- Ask the students to answer the Student Book questions. [O1, 2&3]

Extend

- Ask students able to progress further to **present the data** on each of the composites from Worksheet 3.3.17 and the Student Book and use this to produce a **detailed analysis** of how they compare with each other. [O3]

Plenary suggestions

What am I? The students should ask questions about the type of material you have chosen – but only 'Yes' or 'No' answers may be given. Choose a material from polymer, natural polymer, ceramic, composite or human-made composite. [O1]

Select the best material State some applications that various materials have. Ask the students to select the best material for the job, giving their reasons. [O2]

Ask higher-attaining students to **use models and diagrams to summarise** the structure of each of the different types of materials they have learned about in the last few lessons. [O3]

Answers to Student Book questions

1. Glass and carbon fibres are plentiful in supply and very lightweight; they make a composite much stronger.
2. Plastic is lighter, and cheaper to make; it is easier to make a plastic for your required needs.
3. Carbon fibre is much stronger than fibreglass.
4. *Fibreglass* can be used for making pipes, hot tubs, kayaks and canoes.
   Carbon fibre is also used to make trains; Kevlar is a bulletproof composite with carbon fibre.
5. *Carbon fibre* is the strongest and least dense material, with the highest strength-to-weight ratio. *Fibre composites* are stronger than metals; metals are more dense than composites.
6. Fibreglass consists of fibres of glass moulded into plastic; carbon fibre is made by fibres of carbon in plastic; carbon is much stronger than glass.

Answers to Worksheet 3.3.17

1. Concrete and reinforced concrete have the same composition, in terms of the concrete. They are different because reinforced concrete has steel rods, or rods made from other materials, to make it stronger.
2. summaries should include:
   - what the composite is made from
   - what the uses of the composite are
   - what the properties of the composite are
   - any special features, including ideas about cost.
3. A model of fibreglass should show a plastic resin with fibres of glass evenly dispersed throughout. A model of carbon fibre should show much thicker, stronger fibres, of carbon in a plastic matrix. A cermet model should show a ceramic matrix with metal particles interspersed throughout.

3.18 Checking students' progress

The 'Checking your progress' section in the Student Book indicates the key ideas developed in this chapter and shows how students progress to more complex levels. It is provided to support students in:

- identifying those ideas
- developing a sense of their current level of understanding
- developing a sense of what the next steps in their learning are.

It is designed either to be used at the end of a chapter to support an overall view of the progress, or alternatively during the teaching of the unit. Students can self assess or peer assess using this as a basis.

It would be helpful if students can be encouraged to provide evidence from their understanding or their notes to support their judgments. In some cases it may be useful to explore the difference in the descriptors for a particular idea so that students can see what makes for a 'higher outcome'.

It may be useful with some descriptors to provide examples from the specific work done, such as an experiment undertaken or an explanation developed and recorded. If marking and feedback uses similar ideas and phrases this will enable students to relate specific marking to a more general sense of progress.

To make good progress in understanding science students need to focus on these ideas and skills:

Students who are making modest progress will be able to:	Students who are making good progress will be able to:	Students who are making excellent progress will be able to:
Write word equations to represent the decomposition of metal carbonates.	Use observations from thermal decomposition reactions to make inferences about metal reactivity.	Write balanced symbol equations for the decomposition of metal carbonates.
Give uses of displacement reactions and write word equations to represent them.	Use models to explain displacement and relate it to the reactivity series.	Write balanced symbol equations for displacement reactions.
Describe different ways to extract metal ores from the Earth and describe the associated environmental issues.	Explain how metals are recycled and how this affects the environment.	Evaluate the positive and negative aspects of metal mining and extraction.
Describe the use of carbon in extracting iron from its ore.	Describe the process of extracting iron from its ore in a blast furnace.	Use balanced symbol equations to make predictions about the mass of iron produced when extracted from ore, showing that mass is conserved, and explain the advantages of using carbon.
Write word equations for the reactions between carbon and metal ores.	Describe the extraction processes for lead, copper and zinc.	Work out the yield of an extraction process.
Describe what is meant by the terms exothermic and endothermic reactions, with examples.	Explain the energy changes taking place during an exothermic and an endothermic reaction.	Use energy-level diagrams to compare the energy in the reactants and products of exothermic and endothermic reactions, explaining the energy changes in the particles.
Describe what a catalyst is and give examples.	Interpret data to explain how a catalyst affects a reaction.	Explain how a catalyst works.
Describe what is meant by the term 'ceramic', describing their properties and uses, with some examples.	Explain how different types of ceramic vary in their properties.	Explain how the chemistry and bonding within a ceramic affects its properties.
Describe what is meant by the term 'polymer', using examples of natural and human-made polymers.	Describe the properties of polymers, explaining how these relate to their uses.	Explain how the properties of polymers are affected by their bonding, using simple models.
Describe what is meant by the term 'composite', using examples of natural and human-made composites.	Describe the properties of composites, explaining how the properties relate to their uses.	Use models to explain how composites are constructed and use these to explain their properties.

3.19 Answers to Student Book Questions

This table provides answers to the Questions section at the end of Chapter 3 of the Student Book. It also shows how different questions assess attainment in terms of the focus and style of a question as well as the context. Question level analysis can indicate students' proficiency in approaching different aspects of scientific understanding and different types of answer.

Q	Answer		Focus			Style			Context			
		Marks available	Knowledge & understanding	Application	Evaluation of evidence	Objective test question	Short written answer	Longer written answer	Metals and their extraction from ores	Reactivity series	Energy changes in reactions and catalysts	Polymers, ceramics and composites
1	b	1	x			x						x
2	a	1	x			x				x		
3	c	1	x			x					x	
4	d	1	x			x			x			
5	• a substance that speeds up or slows down a reaction • a substance that is not changed by the chemical reaction	1 1	x x				x x				x x	
6	• a polymer is made up from many repeating units chemically bonded together in long chains • a composite is two or more substances combined but not usually chemically	1 1	x x				x x					x x
7	any **four** from: • hydrogen and oxygen molecules collide • the collision puts energy into the molecules • bond-breaking is endothermic • the reactant molecules break apart into separate atoms • the atoms make new bonds, releasing energy • bond-making is exothermic • there is more energy released when the new bonds are made compared to the energy taken in when the reactant bonds are broken	4	x					x			x	
8	b	1		x		x				x		
9	c	1		x		x				x		
10	a	1		x		x						x
11	c	1		x		x					x	

Q	Answer		Focus			Style			Context			
		Marks available	Knowledge & understanding	Application	Evaluation of evidence	Objective test question	Short written answer	Longer written answer	Metals and their extraction from ores	Reactivity series	Energy changes in reactions and catalysts	Polymers, ceramics and composites
12	• 20 tonnes of zinc could be theoretically extracted • 10 tonnes are actually obtained	1 1		x x			x x		x x			
13	scratching a tin can • tin is less reactive than steel; the oxygen present in the air will react readily with the steel • zinc is more reactive than steel; if it is scratched, more zinc will react readily with the oxygen and protect it further; the zinc is sacrificed for the steel	1 1		x x			x x			x x		
14	any **four** from: • there will be scarring of the landscape • subsidence of the ground may occur, causing areas to collapse and become unsafe • there will be a leachate of copper that could enter rivers and lakes and kill wildlife • waste rock could be dumped, spoiling the view • lots of dust could be formed during blasting • lots of noise from trucks and heavy machinery • carbon dioxide will be emitted from the transport and use of machinery causing global warming	4		x				x	x			
15	• better catalyst: steeper curve • poorer catalyst: less steep curve • they should all follow the same shape • they should all make the same volume of oxygen	1 1 1 1			x x x x			x x x x			x x x x	
16	a) if aluminium is coated over the titanium; because it is more reactive it will form an oxide layer; that will prevent oxygen reaching the titanium b) if copper is coated over the tianium; it will react much more slowly with the air; than the bare titanium would	1 1			x x		x x			x x		
	Total possible:	**30**	**12**	**12**	**6**	**8**	**10**	**12**	**7**	**7**	**12**	**4**

4.1 Introduction

When and how to use these pages

The Introduction in the Student Book indicates some of the ideas and skills in this topic area that students will already have met from KS2 or from previous KS3 work, and provides an indication of what they will be studying in this chapter. *Ideas you have met before* is not intended to comprehensively summarise of all the prior ideas, but rather to point out a few of the key ones and to support the view that scientific understanding is progressive. Even though students might be meeting contexts that are new to them, they can often use existing ideas to start to make sense of them.

In this chapter you will find out indicates some of the new ideas that the chapter will introduce. Again, it isn't a detailed summary of content or even an index page. Its purpose is more to act as a 'trailer' and generate some interest.

The outcomes, then, will be recognition of prior learning that can be built on, and interest in finding out more.

There are a number of ways this can be used. You might, for example:

- Use *Ideas you have met before* as the basis for a revision lesson as you start the first new topic.
- Use *Ideas you have met before* as the centre of spider diagrams, to which students can add examples, experiments they might have done previously or what they found interesting.
- Make a note of any unfamiliar/difficult terms and return to these in the relevant lessons.
- Use *In this chapter you will find out* to ask students questions such as:
  - Why is this important?
  - How could it be used?
  - What might we be doing in this topic?

Overview of the chapter

In this chapter, students will learn about the composition of Earth's atmosphere, how it has evolved and what caused it to change. They will learn how human activities can damage and limit the Earth's resources, including the atmosphere. The students will explore how these changes affect the Earth and all life on it. Natural recycling and the recycling of waste materials are considered. The students will also learn about the structure of the Earth. They will study igneous, sedimentary and metamorphic rocks, and the rock cycle.

This chapter offers a number of opportunities for students to model processes and to evaluate the usefulness of the models. They undertake a range of investigations into the amount of oxygen in the atmosphere, the effect of human activities on plant growth and recycling materials.

Obstacles to learning

Students may need extra guidance with the following:

- **The rock cycle** They may need help in distinguishing between weathering and erosion. They may think that all intrusive rocks cooled more slowly than extrusive rocks.
- **Environmental damage** They may need help in distinguishing between the different effects of air pollution – acid rain, the greenhouse effect and damage to the ozone layer. They may think that the greenhouse effect is caused entirely by humans.
- **Other misconceptions** The following misconceptions are common – the Earth is indestructible; natural resources are unlimited.

	Topic title:	Overarching objectives:
2	Understanding our atmosphere	The composition of the atmosphere
3	Exploring the effects of human activity	The production of carbon dioxide by human activity and the impact on climate
4	Understanding the global warming debate	
5	Understanding how carbon is recycled	The carbon cycle
6	Exploring damage to the Earth's resources	Earth as a source of limited resources and the efficacy of recycling
7	Considering the importance of recycling	
9	Understanding the structure of the Earth	The composition of the Earth The structure of the Earth
10	Exploring igneous rocks	The rock cycle and the formation of igneous, sedimentary and metamorphic rocks
11	Studying sedimentary rocks	
12	Using metamorphic rocks	
13	Understanding the rock cycle	

4.2 Understanding our atmosphere

Lesson overview

Learning objectives

- Describe the composition of our atmosphere.
- Describe how the atmosphere has changed over time.
- Explain why the atmosphere has changed.

Learning outcomes

- Identify the percentages of the different gases that make up the atmosphere. [O1]
- Describe how the percentages of different gases in the atmosphere have changed over time. [O2]
- Explain why the atmosphere has changed in terms of oxygen and carbon dioxide proportions. [O3]

Skills development

- Thinking scientifically: analyse data
- Working scientifically: make predictions
- Learner development: collaborate effectively

Resources needed 12 cm ball; large sheets of paper; glue; materials and equipment as detailed in the Technician's notes; Worksheet 3.4.2 (page 2 copied onto card); Practical sheet 3.4.2; Technician's notes 3.4.2

Digital resources Quick starter; Interactive activity: Drag the statements to the correct group - element in clean air, compound in clean air, or pollutant in air; Interactive activity: Match the components of air to the percentages present

Common misconceptions The Earth's atmosphere has always been as it is now. Oxygen is the main gas in the Earth's atmosphere.

Key vocabulary atmosphere

Teaching and learning

Engage

- **Pair talk** Ask the students to **discuss** 'Why is the Earth's atmosphere important?'. Take feedback and discuss ideas. [O1]
- **Think, pair, share** Show the class a ball – ask them how far the Earth's atmosphere extends. Take feedback and vote on their ideas. (It is about 1 mm for a 12 cm ball, but the breathable atmosphere is only 0.1 mm.) [O1]

  Higher-attaining students may be able to name the main gases in the atmosphere.

Challenge and develop

- **Pair work** The students read the 'What is the composition of our atmosphere?' section in the Student Book. They **discuss** and answer questions 1 and 2. [O1]
- Ask the students to **predict** how the atmosphere in the room will change if it is sealed with everyone inside – they should **justify** their predictions. [O1&3]
- Demonstrate the experiment to measure the amount of oxygen in the air – see Technician's notes 3.4.2. They can use Practical sheet 3.4.2 to **follow the procedure** and **record** their results. They can then **discuss** and answer the questions on the practical sheet. [O1]
- **Group work** Write 'The Earth's atmosphere is just the right mix for life' on the board. Groups **discuss** and **compile a table** with reasons why they agree, disagree or are unsure about the statement. Take feedback and discuss. Lead into the fact that our atmosphere has changed over time. [O1&2]

Explain

- **Group work** The students read the 'How has Earth's atmosphere evolved?' section in the Student Book. They **discuss** and answer questions 3 and 4. Ensure that they understand that early Earth had an atmosphere with little oxygen, high temperatures and UV light levels, and frequent thunderstorms. [O2]
- **Pair work** The students complete task 1 of Worksheet 3.4.2. They will need the cards from page 2 of the worksheet. [O1&2]
- **Pair research activity** Students complete task 2 of Worksheet 3.4.2. They use the internet to **research** the composition of the atmosphere from 4 billion years ago to the present day and **draw a timeline** of the changes. If the internet is not available or time is short, give the students the information about 'How Earth's atmosphere has changed' on page 3 of the worksheet. [O2&3]

Consolidate and apply

- **Pair talk/Pairs to fours** The students read the 'What caused the changes in composition?' section in the Student Book. They **discuss** and answer questions 5 and 6 and complete task 3 of Worksheet 3.4.2. [O3]

Extend

- Ask students able to progress further to work in pairs to **research** information and **design** a poster that describes the different named layers of the atmosphere and the features of these layers.

Plenary suggestions

Summarise Ask the students to work in pairs to summarise their learning from this lesson in no more than 30 words. [O1, 2&3]

Answers to Student Book questions

1. the layer of gases surrounding the Earth
2. The Earth's atmosphere consists mainly of nitrogen (78%). There is 21% oxygen; and small amounts of other gases – for example 0.9% argon, 0.04% carbon dioxide, variable amounts of water vapour, and trace amounts of other gases.
3. initially: CO_2, NH_3, CH_4 and water vapour; there was less water vapour as the Earth cooled; CO_2 and NH_3 dissolved in oceans; less CO_2 as plants photosynthesised and produced O_2; CO_2 and O_2 became stabilised at their current levels
4. they should include the ideas that Earth's crust cooled and became solid; clouds formed; rain fell and the oceans formed; the atmosphere started to change; the conditions very slowly became suitable for the formation of life
5. O_2 increased due to release from increased photosynthesis; CO_2 decreased due to use in photosynthesis; CO_2 also dissolved in oceans and became trapped in rocks/fossil fuels
6. the amount of CO_2 used in photosynthesis equals the amount released in respiration; the amount of O_2 used in respiration equals the amount released in photosynthesis

Answers to Worksheet 3.4.2

1. a) the correct percentage of each gas identified b) H, D, E, B, F, G, C, A
2. initially: CO_2, NH_3, CH_4, H_2 and water vapour; then the Earth cooled and water vapour condensed; CO_2 dissolved in oceans and became trapped in rocks; as plant life evolved and began to photosynthesise, CO_2 was removed from the atmosphere and oxygen was released; increased oxygen allowed Earth to support animal life; combustion releases vast amounts of CO_2 and deforestation means less CO_2 absorbed
3. a) CO_2 rose rapidly, as hydrogen, helium and other gases escaped. It then started to fall as plants evolved and used it in photosynthesis. O_2 started low; there was a sudden rise due to the evolution of plants, which released it in photosynthesis. Evolving animal life caused CO_2 to rise slightly as they released the gas in respiration. Levels eventually stabilised.

Answers to Practical sheet 3.4.2

1. to make the copper react with the air
2. to push the air backwards and forwards over the copper
3. some of the air/oxygen in the air has reacted with the copper
5. Errors in measurements cause variation.
7. a) not all the oxygen has been used up; not enough copper was used; not heated for long enough

   b) the air has not been pushed into one syringe; air is leaking from the joints in the apparatus
8. After the air has been allowed to cool and the volume measured, it should be reheated and the measurement repeated until the final volume remains constant; if there is a leak, the volume will continue to decrease.

4.3 Exploring the effects of human activity

Lesson overview

Learning objectives

- Describe examples of human activity that cause air pollution.
- Explain the effects of smog, acid rain and damage to the ozone layer.

Learning outcomes

- Describe sources of air pollution that affect the atmosphere. [O1]
- Explain how air pollution damages the atmosphere and suggest how we can reduce the effects. [O2]
- Explain new technologies that help in reducing air pollution, and analyse their impact. [O3]

Skills development

- Thinking scientifically: analyse data
- Working scientifically: interpret evidence
- Learner development: collaborate effectively

Resources needed materials and equipment as detailed in the Technician's notes; large sheets of paper and coloured pens; Worksheet 3.4.3a; Worksheet 3.4.3b; Practical sheet 3.4.3; Technician's notes 3.4.3

Digital resources Quick starter; Slideshow: Effects of air pollution - Carbon dioxide, UV, ozone and acid rain; Video

Common misconceptions The Earth is indestructible.

Key vocabulary ozone layer, acid rain, smog

Teaching and learning

Engage

- Ask the students to **brainstorm** human activities that damage the atmosphere. [O1]

Challenge and develop

- **Think, pair, share** The students read the 'What causes air pollution?' section in the Student Book and answer questions 1 and 2. [O1]
- **Pair work** The students carry out task 1 of Worksheet 3.4.3a. They **research** human activities that cause air pollution, using the Student Book, their discussions and other suitable information that you can provide. Give them a time limit to **design** a poster to describe the effects of these activities. [O2]

Explain

- **Jigsaw activity** Arrange the students into 'expert' groups to work together to learn more about one of: smog; acid rain; damage to the ozone layer. You can give them the relevant information sheet from Worksheet 3.4.3b. Then the groups re-form so that there is at least one expert on each topic in each group to **teach** the rest about their topic. [O1&2]

  Assign ozone layer damage to higher-attaining students.

- **Pair talk** The students read and **discuss** the 'What are the effects of air pollution?' section in the Student Book. Draw attention to the word equations for the reaction of limestone with each type of acid rain. They then work on their own to answer questions 3 and 4. [O2]
- **Pair work** The students can then complete task 2 of Worksheet 3.4.3a. They **write** a newspaper article to **explain** how human activities affect the ozone layer and how we can reduce these effects. [O2]

Consolidate and apply

- **Group work** The students **set up an investigation** concerning the effect of acid rain on cress germination, following Practical sheet 3.4.3. Each group will test one type of acid rain (sulfuric acid, nitric acid or carbonated mineral water – see Technician's notes 3.4.3) and class results will be collated. Seeds will need to be left to germinate for a week. Allow time in a later lesson for the analysis. Discuss the effect of each type of acid rain and conclude which had the greatest impact on germination. [O2]

Extend

Ask students able to progress further to:

- work in pairs to **design a presentation** to **explain** how human activities damage the atmosphere and to **analyse** the effect of technological advances on the quality of the atmosphere (task 3 of Worksheet 3.4.3a). [O3]
- attempt to **write the symbol equations** for the production of acid rain. ($CO_2 + H_2O \rightarrow H_2CO_3$; $SO_2 + H_2O \rightarrow H_2SO_3$; $2NO_2 + H_2O \rightarrow HNO_2 + HNO_3$)

Plenary suggestions

How, where, when, why and what? In groups, the students **write questions** on this topic starting with 'How', 'Where', 'When', 'Why' and 'What' along with answers. They then **answer** another group's questions and **peer-assess** them. Take feedback. [O1, 2&3]

Answers to Student Book questions

1. unwanted waste substances that damage the Earth and the atmosphere
2. they should describe forest fires; volcanic eruptions; wind erosion; pollen dispersal; natural radioactivity
3. When gases (such as sulfur dioxide) are released in burning, they dissolve in water vapour in the atmosphere.
4. The ozone layer helps in protecting us from harmful rays from the Sun that can cause cancer.
5. if there had been technological advances that we have had since, there would have been sulfur-less fuel; hydrogen engines; catalytic converters; filters/scrubbers/electrostatic precipitators in chimneys; renewable energy technology (no emissions)
6. they should include a discussion of the feasibility of wind farms; solar power; hydroelectric power; biofuels; with at least one advantage and disadvantage of each; e.g. wind farms – plenty of strong wind in exposed areas of the UK, but cannot use if the wind is too strong/weak

Answers to Worksheet 3.4.3a

1. Check that the posters include correct examples, including those in the Student Book.
2. Check that the articles include correct examples, including those in the Student Book.
3. Check that the presentations include all of the above, and also an analysis of the effectiveness of each strategy in improving the quality of the atmosphere, plus acknowledgement that all countries need to adopt these pollution-reducing technologies.

Answers to Practical sheet 3.4.3

1. yes; the same number of seeds was used for each test; the same volume of solution was added; the same sized dishes were used; and the seeds were all left for the same time
2. the more concentrated the acid, the fewer seeds germinated
3. to stop the evaporation of water; which could have affected the results
4. it is a control, to show that the seeds will grow normally given the right conditions (to show that there is nothing wrong with the seeds)
5. All groups should find that as acid rain concentration increases, germination rate decreases. Carbon dioxide in rainwater (sparkling water) will have the least effect on seed germination, and nitrogen oxide in rainwater should have the greatest effect (least seeds germinate).

4.4 Understanding the global warming debate

Lesson overview

Learning objectives

- Describe the effects of global warming.
- Explain the consequences of global warming for living things.
- Evaluate the arguments for human activity impacting on global warming.

Learning outcomes

- Define global warming and describe how it warms up the atmosphere. [O1]
- Explain how global warming affects different ecosystems. [O2]
- Identify that the greenhouse effect is natural and that the impact of human activity on it is still debated. [O3]

Skills development

- Thinking scientifically: ask questions
- Working scientifically: develop explanations
- Learner development: respect others

Resources needed Worksheet 3.4.4 (with the page of 'cards' copied onto card)

Digital resources Quick starter; Interactive activity: Reorder the sentences to describe how we receive energy from the Sun; Interactive activity: Match the sea surface temperature to the year

Common misconceptions The greenhouse effect and the depletion of the ozone layer are the same thing.

Key vocabulary greenhouse effect, global warming

Teaching and learning

Engage

- **Pair talk** Write the following on the board: 'greenhouse effect', 'atmosphere', 'greenhouse gas', 'carbon dioxide', 'carbon emissions', 'deforestation', 'global warming'. The students try to **explain** these words to a partner. Take feedback. [O1]
- **Think, pair, share** Ask the students to write down answers, individually, to 'What is global warming?' They then **share** their definitions and **compare** similarities and differences. They **combine** their definitions into a new, more complete or accurate one (task 1a of Worksheet 3.4.4). Take feedback from the class and **agree on** a class definition. [O1]

Challenge and develop

- The students read the 'What is global warming?' section in the Student Book and answer questions 1 and 2. [O1]
- **Pairs to fours** They complete the **sequencing activity** on the greenhouse effect (task 1b of Worksheet 3.4.4). They then **compare** answers with those of another pair and **come to an agreement** on the correct order. Take feedback. Reinforce the fact that the greenhouse effect is a natural phenomenon (without which the Earth would be too cold for life). [O1]
- Discuss the term 'enhanced greenhouse effect', referring to Figure 3.4.4a in the Student Book. Ensure that the students grasp that it is this that is believed to be the cause of global warming. [O1]

Explain

- **Pair talk** The students read and **discuss**, in pairs or small groups, the 'Impact of global warming' section in the Student Book and answer questions 3 and 4. [O2]

- They can then **analyse** two graphs of global warming data (task 2 of Worksheet 3.4.4). [O2]
- The students read and **discuss** the 'Are we to blame?' section in the Student Book and answer questions 5 and 6. [O3]

Higher-attaining students may be able to **predict** how the world's climate could change over the next 100 years if greenhouse gas emissions are not reduced, and the effects such climate changes could have on humans.

Consolidate and apply

- **Pair work** Ask the students to do some further **research**, and **design** a poster to **explain** the consequences of global warming for living things. [O2]
- They **identify arguments** *for* human-produced global warming, arguments *against* and facts that are *agreed* by both sides. They then **work individually** to complete a **piece of extended writing** giving a balanced argument and conclusion about the cause(s) of global warming. Task 3 of Worksheet 3.4.4 guides them through this. [O3]

Extend

- If students are able to progress further, set up a class **debate** – divide the class into halves. Each half nominates a proposer and a seconder, and then identifies arguments for and against the statement 'Human activities have caused the rate of global warming to increase'. For each argument, they decide on a counter-argument. Tell them whether they will debate for or against the motion, and start the debate. [O3]

Plenary suggestions

Summarise The students write a short paragraph that **explains** the link between the greenhouse effect and global warming. [O1, 2&3]

Answers to Student Book questions

1. The Sun's radiation passes through the atmosphere, is absorbed by the Earth's surface and is re-radiated as heat energy. This heat energy is absorbed or reflected by the greenhouse gases, instead of escaping into space. This causes the atmosphere to retain its warmth, and so to support life on Earth.
2. *carbon dioxide* (plant decay, volcanic eruptions, product of respiration); *methane* (cows, pigs, landfill, rice fields etc.); *nitrous oxide* (fuel combustion and industry); *CFCs* (aerosols, solvents in cleaners, coolants in refrigeration and air conditioning); *water vapour* (respiration, any correct reference to water cycle); *ozone* (naturally occurring and from incomplete combustion of fuels)
3. *polar bears*: polar ice melts – loss of habitat; *penguins*: polar ice melts – loss of habitat; *mosquitoes*: countries get warmer – increased habitat
4. low-lying land would be submerged, with the loss of agricultural land; desertification would cause loss of agricultural land; agricultural pests may thrive in warmer climates; all these reduce crop volume and damage food security; but possible faster crop growth and longer growing seasons may compensate
5. The rate at which the Earth is heating has increased dramatically; this rate not been seen in the previous 10 000 years.
6. *The Earth's sea surface temperature* has been below the 2500-year average over the last 500 years or so, although it has been steadily increasing for the last 300 years or so; no conclusive evidence that the Earth is heating up dramatically.

   The Arctic melt season throughout 1990–2010 has varied considerably; it peaked in 1997 at about 218 days; in 2003 it was down to about 167 days; in 2010 it was back up to 185 days; no conclusive evidence of increased polar ice cap melting due to global warming.

Answers to Worksheet 3.4.4

1. a) H; D; G; C; A ; E ; F; B b) the definition should include the idea that it is the rise in the average temperature of the Earth, because of an increase in the greenhouse effect due to increased amounts of grenhouse gases in the Earth's atmosphere
2. a) it is rising b) about 0.85 °C c) it is rising steeply, especially since around 1900

   d) there may be a relationship; as the amount of CO_2 increases, the global average temperature rises
3. a) as a minimum: current rate of change not seen in last 10 000 years; research shows that human activity (e.g. increased combustion of fossil fuels) is increasing global warming; but sea surface temperature over past 3000 years has changed very little; summer melt season in 2010 was about the same as in 1990

   b) *balanced argument* to include: significant scientific evidence points to human activities (e.g. burning fossil fuels) being mainly responsible for the current increased levels of greenhouse gases in the atmosphere; there are also natural causes, such as volcanic eruptions; other factors: poor land use; waste disposal; use/ disposal of CFCs; *discussion of balance* between natural cycle and human activities; the possibility of another, unknown, cause; *reasoned conclusion* based on the evidence discussed and suggestions for sensible action to be taken

4.5 Understanding how carbon is recycled

Lesson overview

Learning objectives

- Describe the carbon cycle.
- Explain how human activity increases the amount of carbon in the atmosphere.
- Explain what is meant by a 'carbon footprint'.

Learning outcomes

- Use the carbon cycle to describe ways in which carbon enters and leaves the atmosphere. [O1]
- Explain how each of the stages in the carbon cycle affects the amount of carbon in the atmosphere. [O2]
- Define 'carbon footprint' and analyse the carbon footprint of different scenarios. [O3]

Skills development

- Thinking scientifically: analyse data
- Working scientifically: interpret evidence
- Learner development: plan progress

Resources needed (optional) materials and equipment as detailed on the Practical sheet; Worksheet 3.4.5; Practical sheet 3.4.5

Digital resources Quick starter; Interactive activity: Drag the statements to the correct group - decomposers or green plants; Slideshow: Carbon cycle - releasing carbon dioxide; Interactive activity: Match the changes in the carbon cycle to the correct term

Common misconceptions There is no debate about whether or not increased levels of greenhouse gases will warm the planet. The only uncertainty lies in how much and how quickly the planet will warm.

Key vocabulary carbon footprint

Teaching and learning

Engage

- The students **brainstorm** to trace the possible movement of a carbon atom around the globe. Record ideas on the board to form a simplified carbon cycle, including at least eating and respiration in animals and photosynthesis in plants. Ask them to **think** about oceans, rocks and minerals – do they contain carbon? [O1]
- **Pairs** Using the ideas on the board, the students **illustrate** their own version of the (natural) carbon cycle. You might want to give them the option of building a carbon cycle using 'props' such as toy animals, plants, shells and so on – see Practical sheet 3.4.5. [O1]

Challenge and develop

- **Back-to-back** In their pairs from the previous activity, the students exchange carbon cycle illustrations with another pair. Each pair of students (A and B) then sit back-to-back. Student A **describes** the diagram to student B (who cannot see it). Student B **asks questions** to clarify their understanding and **draws** an accurate copy of the diagram. After 5 minutes they **compare** the original diagram and student B's drawing. [O1]
- **Think, pair, share** The students read the 'Natural recycling' section in the Student Book and then **agree on** answers to questions 1 and 2. [O1]

  Higher-attaining students can **write a symbol equation** for question 1.
- **Pair work** The students complete task 1 on Worksheet 3.4.5 to **identify** ways in which carbon enters and leaves the atmosphere. [O1]

Explain

- **Think, pair, share** They read the 'Explaining the carbon cycle' section in the Student Book and then **agree on** answers for questions 3 and 4. [O2]
- **Pair work** Use the illustrated carbon cycles that students created earlier in the lesson. They use different colours to add or remove processes or add annotation so that it **compares** the carbon cycle that takes place today with that of 1000 years ago. [O2]
- **Think, pair, share** The students read the 'Carbon footprint' section in the Student Book and **agree on** answers for questions 5 and 6. They can **calculate** their own footprint using an online calculator and **suggest** how they could reduce this. [O3]

Consolidate and apply

- In groups of four, the students **act out** the carbon cycle using paper signs/mini-whiteboards to help them. One person acts as a narrator. They then **peer-assess** the presentations. [O1&2]
- **Independent work** Students should complete tasks 2 and 3 of Worksheet 3.4.5 on their own. [O2&3]

Extend

- Ask students able to progress further to **research**:
  - what different countries can do to reduce the impact of global warming (in the short term and long term)
  - how actions that are taken will affect different groups of people in different countries.

Plenary suggestions

Summarise The students work in pairs to summarise the lesson in no more than 30 words. Ask some pairs to **share** their summaries and ask the class to **peer-assess** using 'three stars and a wish'. [O1, 2&3]

Answers to Student Book questions

1. carbon dioxide + water → glucose + oxygen; carbon from the CO_2 in air is used by plants to produce glucose molecules
2. they should include the idea that decomposers return carbon to the atmosphere; otherwise it would be trapped in dead organisms
3. any four of: weathering of limestone rock; decomposing organisms; decomposing organic waste; respiration; volcanic activity; combustion
4. photosynthesis
5. they should indicate that it is a measurement of the carbon impact that all of a person's activities have on the environment (including equipment they use)
6. developed countries have more industry; require more electricity production; produce more transport fumes; consume more products; etc.

Answers to Worksheet 3.4.5

1. a) B; D; E; G; L b) A; C; F; H; I; J; K
2. a) *increase* CO_2 in the atmosphere: respiration; volcanic eruptions; action of decomposers; combustion; weathering of limestone rock; *decrease* CO_2 in the atmosphere: photosynthesis

   b) *the amount of CO_2 removed* from the atmosphere by photosynthesis; diffusion of CO_2 into lakes/oceans; formation of limestone rock; and fossilisation; *was approximately equal to the amount of CO_2 released*; by respiration; volcanic eruptions; action of decomposers; combustion; and weathering of limestone rock

3 a) more industrialisation; increased use of transport, including international travel; more importing/exporting of food and other goods; greater production of electricity; etc.

   b) *Rani* has a very small carbon footprint and uses very little electricity, walks to school, etc; *Ben* has a very large carbon footprint that could be reduced by watching less TV, using the computer less, having one holiday in the UK, etc.; *Timur* has a larger footprint than *Rani* because his family have more electrical appliances etc., but he cycles to school; however, his footprint is very low compared to *Ben's*; *Timur* could reduce his footprint by having a shower instead of a bath, watching less TV, etc.

4.6 Exploring damage to the Earth's resources

Lesson overview

Learning objectives

- Describe resources that the Earth provides.
- Explain how human activity limits these resources.
- Justify decisions about making changes to the environment.

Learning outcomes

- Name some natural resources (such as wood, rocks, air, water) that the Earth provides and classify them as renewable or non-renewable. [O1]
- Explain how human activities affect the Earth's resources; describe examples, such as overfishing or open-cast mining. [O2]
- Analyse the impact of a development on the environment and justify decisions regarding it. [O3]

Skills development

- Thinking scientifically: ask questions
- Working scientifically: present evidence
- Learner development: act responsibly

Resources needed Worksheet 3.4.6

Digital resources Quick starter; Interactive activity: Drag the statements to the most appropriate group - scars the landscape, pollutes the air, or damages or destroys natural habitats; Video

Common misconceptions The Earth has an unlimited supply of natural resources. Non-renewable resources can be remade.

Key vocabulary renewable resource, non-renewable resource, sustainable

Teaching and learning

Engage

- **Fours** Ask the students to **brainstorm** what resources we need to survive. Take answers from the class, list them on the board and **discuss** them as a class. [O1]

Challenge and develop

- Ask the students to read 'The Earth's vital resources' section in the Student Book and answer questions 1 and 2. [O1]
- **Pair work** Ask the students to do task 1 of Worksheet 3.4.6, **producing a presentation** about resources that the Earth provides and how we use them. Ask selected pairs to give their presentation and ask the class to **peer-assess**. [O1]

Explain

- **Group work** The students read the 'Damaging the planet' section in the Student Book and **discuss** why some areas of the environment are not as vulnerable to human activity as others – for example, the deep ocean and the polar ice caps. [O2]
- **Independent work** Ask the students to answer questions 3 and 4 on their own. [O2]
- **Group work** Refer to the natural resources identified by the students in answer to question 1 in the Student Book. Ask the students, in their groups, to **identify** how human activity to obtain these resources either limits them or damages the environment. Ask them to **research** how certain human activities damage the

environment – deforestation, landfill, fertilisers/sewage and overfishing/agriculture (using task 2 of Worksheet 3.4.6). [O2]

- The students can then read and **discuss** in their groups the case study in the 'Making sustainable decisions' section of the Student Book, and answer questions 5 and 6. [O3]

  Higher-attaining students could carry out more **research**, and **role-play** their plan to manage the Amazon's resources sustainably.

Consolidate and apply

- **Fours** Ask the students to **analyse** the arguments in the Minestown dilemma activity (using task 3 of Worksheet 3.4.6) to **decide on** a recommendation and to **justify** it. Groups swap reports and **peer-assess** them. [O3]

Extend

- Ask students able to progress further to **research** information and **write** a newspaper article on the Great Pacific Gyre, the world's largest landfill, which is located in the middle of the Pacific Ocean. [O2]

Plenary suggestions

How, where, when, why and what? The students work in groups to **develop questions** on this topic, along with answers. They **answer** one other group's questions and **peer-assess** them. Take feedback. [O1, 2&3]

Answers to Student Book questions

1. any six named resources; and their uses – e.g. *water*: for drinking, cooling power plants, flushing toilets etc.
2. correct classification depending on the named resources in question 1
3. some are high-demand resources, e.g. copper and diamond mines in Africa; and are more easily accessible
4. the mining process has polluted the stream; causing a change in habitat for the animals and plants; living in and around the water; they may die; animals coming to drink the water may also be affected; it looks unpleasant
5. deforestation, changes to soil quality and flooding the land have caused a loss of habitat; for many animals and plants that lived here
6. replant trees; only cut down the number of trees that will been replanted; use traditional methods to harvest timber; rather than modern mechanised methods; build limited reservoirs in other areas that have infertile soils; any other sensible responses

Answers to Worksheet 3.4.6

2. a) *deforestation*: loss of biodiversity and habitat; results in infertile soils; etc.; *landfill*: loss of biodiversity and habitat; polluted soil; etc.; *fertiliser/sewage*: water may not be suitable for human/wildlife consumption; loss of biodiversity and suitable habitat; eutrophication; etc.; *overfishing*; loss of species; etc.; *agriculture*: infertile soils reduce the numbers of producers; overgrazing destroys pastureland; monocultures are more vulnerable to disease; loss of hedgerows causes loss of habitat and biodiversity; etc.
3. a) how much farming land would be lost; the current volume and speed of traffic that is just passing through the town and will not stop; current road damage; times and occurrences of congestion; the predicted costs of the by-pass; the endangered species that would be affected; the possibility of building a new protected site for these rather than re-routing the by-pass

   b) Possible arguments *for* the by-pass: it will reduce congestion, noise and pollution on the current roads; there will be less damage to roads from heavy wagons; more industry may be attracted to the area if there are better roads, creating jobs.

   Against: cost to the taxpayer; damage to farmland; damage to leisure facilities (golf course); environmental damage to the whole area.

4.7 Considering the importance of recycling

Lesson overview

Learning objectives

- Describe examples of recycling.
- Explain the benefits and limitations of recycling schemes.
- Compare the efficiency of recycling methods.

Learning outcomes

- Name some materials that can be recycled and describe the benefits. [O1]
- Explain how factors such as cost, pollution, energy needs etc. limit the efficiency of recycling schemes. [O2]
- Compare the efficiency of recycling glass, paper, aluminium and plastic. [O3]

Skills development

- Thinking scientifically: analyse data
- Working scientifically: develop explanations
- Learner development: communicate effectively

Resources needed materials and equipment as detailed in the Technician's notes; large sheets of paper and coloured pens; Worksheet 3.4.7; Practical sheet 3.4.7; Technician's notes 3.4.7

Digital resources Quick starter; Interactive activity: Place the materials in order, from the fastest to decompose to the slowest; Slideshow: Recycling issues - A look at some issues of recycling, including its problems; Hangman: Key vocabulary game

Key vocabulary landfill, down-cycle

Teaching and learning

Engage

- **Brainstorm** Ask 'What is recycling and why is it important?' Take feedback from the class and agree on a definition of recycling. Discuss why recycling is important, building on the previous lesson's learning. [O1]

Challenge and develop

- **Think, pair, share** The students **identify** materials that can be recycled. Take answers from the class and list their ideas on the board. They read the 'Why recycle?' section in the Student Book and answer questions 1–3. [O1]
- **Fours** The students **discuss** the advantages and disadvantages of recycling materials. Each group **presents** their ideas to the class. Remind them of the 'Why recycle?' section to help them. [O1&2]

Explain

- The students **design** a poster to **describe** types of materials that can be recycled and the benefits of recycling using task 1 of Worksheet 3.4.7. [O1]
- Students read the 'Limitations of recycling' section in the Student Book and answer questions 4 and 5. [O2]
- Ask them to complete the 'Looking more closely at recycling' activity using task 2a on Worksheet 3.4.7. [O2]
- If possible, demonstrate the recycling of aluminium – see Technician's notes 3.4.7 for the procedure. The students **calculate** the percentage of aluminium reclaimed. They can use Practical sheet 3.4.7, which also has some questions. [O1, 2 &3]

Consolidate and apply

- Ask the students to read the 'Making sustainable decisions' section in the Student Book and answer questions 6 and 7. [O3]
- **Fours** Each student **researches** one of: paper, glass, aluminium or plastic recycling. Their research should include the process, and the advantages and disadvantages compared to using natural resources using task 2b on Worksheet 3.4.7. They then **teach** one another about their material. [O1, 2&3]
- The students **work independently** to complete task 3 of Worksheet 3.4.7 using the research from the previous task. [O1, 2&3]

Extend

For students able to progress further:

- Ask them to **present an argument** for and against the use of plastic packaging and containers. They must include the problems they cause, possible solutions (including alternative forms of packaging) and why plastic has been used. [O1, 2&3]
- Alternatively ask the students to **research** how compost heaps work and give their advantages and disadvantages. [O1]

Plenary suggestions

Summarise Ask the students to work in pairs to summarise the lesson in no more than 30 words. Take feedback and ask them to **peer-assess** the statements. [O1, 2&3]

Two stars and a wish Alternatively the students can work individually to complete 'two stars and a wish' – two 'stars' are things they have done well, learned, understood or enjoyed; a 'wish' is something they could improve, need to complete or need help with. [O1, 2&3]

Answers to Student Book questions

1. for example: paper; plastic; aluminium; glass; copper; rubber; textiles; batteries; car phones; wood; plant material
2. correct bar chart using data in Table 3.4.7a
3. the card used is coated in plastic
4. it conserves the Earth's natural resources for future generations
5. the waste material is mixed up with decaying food waste etc.; there are sharp edges on tin cans, broken glass etc.; some of the waste may be toxic, flammable or otherwise hazardous
6. recycling uses 96% less energy than starting from the natural ore; the ore is scarce; the product is of a good quality and can itself be recycled again
7. there are insufficient trees/wood resources of their own to provide the paper products they need

Answers to Worksheet 3.4.7

1. a) see answer to Student Book question 1.

   b) posters should include some of the following ideas: reduced demand for natural resources; reduced use of fossil fuel for transport and processing; less pollution; release of less greenhouse gas from burning waste; less waste sent to landfill.
2. a) i) 2000/01 ii) compost; 3 700 000 tonnes iii) not an efficient process/some reused in the home/possibly not as much glass waste as other types of waste iv) no collection/they were still refining the process

   b) some possible *advantages* are: saving natural resources; less energy used to recycle than extract new raw materials; less waste sent to landfill so less sites required and less land pollution from leachates

   some possible *disadvantages* are: produce a large amount of pollution; sites are often unsafe and unhygienic; initial setup costs can be very high; some processes use a lot of energy; separation of useful material from waste can be difficult; recycled products are often not high-quality (down-cycled) or durable
3. a) any reasonable comparison based on Table 3.4.7b in the Student Book and their own research
   b) it will help to conserve natural resources for future use; there will still be energy savings even if these are low; provides jobs; reduces waste sent to landfill; reduces greenhouse gas production from burning waste; etc.

Answers to Practical sheet 3.4.7

1. answer will depend on practical result, but should be calculated according to the equation given on the Practical sheet
2. to allow alum crystals to form
3. so that it is at the same temperature as the contents of the beaker; to prevent crystals from re-dissolving

4.8 Applying key ideas

Washable-nappy service

Objectives

- Extract ideas about damage to the Earth's resources caused by landfill sites and the importance of recycling.
- Apply ideas about recycling in nature and by humans to a new situation.
- Use ideas and information about global warming.

Outcome

- Clear and effective responses to questions, indicating understanding and the next steps in learning.

The purpose of this activity is to provide an opportunity to see how successfully students are grasping the key ideas so far. It isn't designed to be used as a formal test – it might be that the students work on the questions collectively. It does provide an opportunity for the teacher to look at written work, engage students in discussions and form ideas about progress being made.

The tasks are progressive. Lower-attaining students should be able to tackle the first two or three tasks and higher-attaining students will find the later ones more challenging.

Resources needed disposable nappy; balance; terry nappy; biodegradable nappy liner; large sheets of paper; highlighter pens; Worksheet 3.4.8

Teaching and learning

Engage

- Challenge the students to **reflect back** over the last few lessons and **identify** some of the key ideas they have learned about. Draw out responses about these ideas and stress that it is important as learners that they develop a sense of their own progress.
- Introduce the context by asking the students if anyone lives with a baby at home. If they do, what kind of nappies do they have? Do they know why their parents chose this type of nappy? Pass round an example of a disposable nappy, a terry nappy and a biodegradable nappy liner. Say that most babies have disposable nappies and ask the student groups to **estimate** the number of nappies a baby wears in 24 hours. They can then **calculate** the number worn in a week and in a year. Weigh a nappy. What weight of used nappies would be sent to landfill in a year per baby? Are they surprised at this number?

  Some students may need to be supported through the calculations.

Challenge and develop

- Give each group a key idea from the topic as a whole, and a large sheet of paper. Ask them to **summarise** everything they can remember about this key idea in 30 seconds. Each group then passes their work on to the next group; repeat the process three or four times. Return the work to the original group and give them time to read it through. Each group then **presents** their ideas to the class. Encourage students to **ask questions** and **discuss** the ideas.

Explain

Ask the students to read the 'Washable nappy service' passage in the Student Book. Depending on their ability to assimilate text, you may need to adopt strategies such as:

- Have the students working in pairs, with copies of the text (supplied as Worksheet 3.4.8) and highlighters. One student highlights key ideas and the other (using a different colour) highlights any words they don't understand.
- Using the text as a source of ideas, students draw a spider chart with 'Washable nappy service' at the centre and the facts, figures, advantages and disadvantages around it.

  Work with lower-attaining students to support them in decoding the text and accessing the ideas.

Consolidate and apply

- Ask students to attempt the tasks in the Student Book, individually or collaboratively. Encourage them to **record** their ideas. Ask them to **present** their responses, either orally or by displaying their work.

Extend

- For some students the later tasks (5 and 6) give opportunities for extension work.

Plenary

How did you do? Ask the students for responses to questions such as 'How well were you able to approach the tasks? Did you manage to present some high-quality responses? Are there some ideas that you need to strengthen in the later part of this topic?'. (If there are, make a note.) Provide the class with some overall feedback, indicating ideas that they have grasped effectively and those they may need to develop further.

Likely responses and next steps in learning

1. This task gives the students an opportunity to recall learning experiences or refer to images and text in the book, and to recall or research the damage that human actions can do to the environment. They should be able to explain that disposable nappies take up a lot of space in landfill and can take hundreds of years to fully break down. Not only do landfill sites look ugly and smell unpleasant, but the toxic material from the nappies is washed into the soil and waterways, and can kill other organisms or harm the environment.
2. This task involves recalling that reusing materials reduces the waste sent to landfill and can help the conservation of environments. Using reusable nappies reduces the use of fossil fuels for the transport to landfill sites and processing of disposable nappies, so less greenhouse gas is produced. Disposable nappies contaminate soil and water as they degrade. Reusable nappies are made from natural fibres and do not expose the baby's skin to chemicals, so are less likely to irritate the skin and cause allergic reactions. If they are washed below 60 °C in an efficient machine and dried outside the carbon emissions are further limited.
3. This task involves explaining parts of the carbon cycle. Students should recall that decomposers (bacteria and fungi) return carbon dioxide to the air as they break down decaying waste matter. The carbon dioxide released is used by plants to photosynthesise and produce food. Plants are the producers in all food chains. During photosynthesis the plants also release oxygen, which is needed by all living things for respiration.
4. This task involves students applying what they have learned about recycling and how human activities can affect the environment. The students should identify the following advantages of recycling nappies: waste material is broken down with other sewage at treatment plants; less waste is sent to landfill; there is less contamination of soil and water; they are made from natural cotton, which can be grown repeatedly and have a very long life span. They should identify the following disadvantages of using disposable nappies: some materials used in the manufacture of disposable nappies are derived from crude oil and the manufacturing process usually involves the use of coal to provide the energy needed. These fossil fuels are in increasingly limited supplies. The materials in most disposable nappies take a long time to degrade in landfill sites.
5. The students need to explore the advantages and disadvantages of sending disposable nappies to landfill AND the advantages and disadvantages of the recycling service in this task. They should use the evidence provided and reach the conclusion that if 'A'-rated appliances are used on full loads, water temperature is kept to 60 °C maximum and washing is dried outside, recycling cloth nappies is the better option for the environment.
6. This task involves the students explaining the concept of global warming (an enhancement of the greenhouse effect, because of additional levels of greenhouse gases). Then they must think about how the energy used when reusing nappies (in the production of laundry products and packaging, transport usage, electricity for machines and so on) can produce greenhouse gases and the impact this may have. Possible ideas include: too much heat is trapped; the Earth's temperature rises above normal; polar ice sheets may melt; sea levels may rise; low-lying land may become submerged; some regions may suffer drought, food shortages and desertification; the growing seasons in some regions may become longer and crop growth may be faster; agricultural pests and tropical diseases may thrive. If, on the other hand, nappy washing is done in 'A'-rated machines on full loads using biodegradable, unfragranced laundry products, washing is done at below 60 °C and nappies are dried outside, the impact may be reduced to less than that of using disposable nappies.

4.9 Understanding the structure of the Earth

Lesson overview

Learning objectives

- Describe the layers of the Earth.
- Describe the characteristics of the different layers.
- Explain how volcanoes change the Earth.

Learning outcomes

- Describe the structure of the Earth and recall that the Earth's surface is constantly changing. [O1]
- Describe the characteristics of each layer of the Earth. [O2]
- Explain how the movement of tectonic plates and volcanoes change the Earth's surface. [O3]

Skills development

- Thinking scientifically: ask questions
- Working scientifically: develop explanations
- Learner development: act responsibly

Resources needed boiled eggs or Cadbury *Creme Eggs*; peaches; *Maltesers*; modelling clay; Worksheet 3.4.9a; Worksheet 3.4.9b (copied onto card)

Digital resources Quick starter; Interactive activity: Drag the statements to the correct group - inner core, outer core, mantle or crust; Video; Interactive activity: Reorder the sentences to describe how volcanoes form and erupt

Key vocabulary mantle, crust, lithosphere, tectonic plate, magma, lava

Teaching and learning

Engage

- **Ideas hothouse** Student pairs **discuss** and **list** what they know about the structure of the Earth. Ask the pairs to form fours and then eights to **discuss further** and come up with an **agreed** list. One student in each group of eight **reports back** to the rest of the class. [O1, 2&3]

Challenge and develop

- **Pairs to fours** The students read the 'Earth's layers' section in the Student Book and **discuss** the answers to questions 1 and 2. Take feedback. Ensure that they understand that in order to float, the plates must be less dense than the mantle. [O1]
- **Pair work** The students complete task 1 of Worksheet 3.4.9a. [O1]
- Place two copies of the information cards from Worksheet 3.4.9b around the room. Continuing in their pairs, one student plays the role of a researcher and the other a scribe. They must **assimilate** the information and **record** it in a suitable format – for example bullet points or a table. [O1&2]

Explain

- The students read the 'Features of the layers' section in the Student Book and answer questions 3 and 4. [O2]
- **Group work** Students **construct a model** of the Earth using different colours of modelling clay. They then **evaluate** a range of other models of the Earth's structure – for example a boiled egg, a peach and a *Malteser* using task 2 of Worksheet 3.4.9a. Explain that each of the items is like a model of the layers of the Earth. Take feedback from the class and discuss. [O1&2]

- Show a suitable video of a volcanic eruption, and/or use recent news articles. The students then read the 'Changing the Earth's surface' section of the Student Book and answer questions 5 and 6. [O3]

Consolidate and apply

- **Group work** Ask the students to complete task 3 of Worksheet 3.4.9a. They **write** a two-minute bulletin for a main news programme on a TV channel, **explaining** how tectonic plate movements and volcanic eruptions change the Earth's surface. They should then **present** their bulletins, which can be **peer-assessed** using 'three stars and a wish'. [O3]

Extend

- Ask students able to progress further to **research** the work of Alfred Wegener and **summarise** his ideas in no more than 30 words. [O1&3]

Plenary suggestions

Spider diagram The students can draw a spider diagram to connect all the information they have learned in the lesson. [O1, 2&3]

Answers to Student Book questions

1. the crust and outer part of the mantle
2. they are less dense than the semi-liquid mantle
3. *oceanic crust*: thinner; denser rock, e.g. basalt; found under the ocean floor; *continental crust*: thicker; less dense rock, e.g. granite; found under land masses
4. the crust is broken up into plates that float on the mantle and can move very slowly, away from or towards each other
5. magma rises and breaks through weak areas in the crust
6. *magma* is molten rock under the Earth's surface; *lava* is molten rock above the Earth's surface

Answers to Worksheet 3.4.9a

1. A – atmosphere; B – inner core; C – outer core; D – crust; E – mantle
2. for example:

Model	How is this model similar to the Earth's structure?	How is this model different from the Earth's structure?
Malteser	*Crust*: (chocolate coating) it is the thinnest layer; it is on the outside *Mantle*: central area *Outer core*: not present *Inner core*: not present	*Crust*: it is soft; not made of rock; smaller; does not crack *Mantle*: solid with air bubbles *Outer core*: not present *Inner core*: not present
Boiled egg	*Crust*: (shell) it is the thinnest layer; it is on the outside; it is rigid or hard; it cracks when pressed on *Mantle*: (egg white) mostly solid; flexible *Outer core*: (yolk) it is in the centre; solid *Inner core*: not present	*Crust*: it is not made of rock; it is smaller than the real thing *Mantle*: it is white; it is the largest layer *Outer core*: yolk is not divided into two layers; not the largest layer; it is yellow *Inner core*: not present
Peach	*Crust*: (skin) it is the thinnest layer; it is on the outside *Mantle*: mostly solid, flexible *Outer core*: the layer is present *Inner core*: the seed inside the stone is solid	*Crust*: it is not made of rock; it is smaller than the real thing *Mantle*: largest layer *Outer core*: solid layer *Inner core*: not made from metal
Modelling-clay model	This will depend on the students' models	This will depend on the students' models

4.10 Exploring igneous rocks

Lesson overview

Learning objectives

- Describe how igneous rocks are formed.
- Explain how the pH of the magma affects the formation of rocks.
- Investigate the effect of cooling rate on the formation of crystals.

Learning outcomes

- Describe how igneous rocks are formed from molten magma and lava and give examples. [O1]
- Explain the relationship between volcano shape, magma pH/ viscosity and rock formation. [O2]
- Explain that slow cooling causes large crystals and fast cooling causes small crystals. [O3]

Skills development

- Thinking scientifically: ask questions
- Working scientifically: develop explanations
- Learner development: collaborate effectively

Resources needed materials and equipment as detailed in the Technician's notes; graph paper; large sheets of paper and coloured pens; Worksheet 3.4.10; Practical sheet 3.4.10a; Practical sheet 3.4.10b; Technician's notes 3.4.10

Digital resources Quick starter; Interactive activity: Drag the statements to the correct group - volcanoes producing acidic magma or alkaline magma; Interactive activity: Match the cooling rate to the crystal size

Common misconceptions Intrusive rocks all cooled more slowly than extrusive rocks.

Key vocabulary igneous, extrusive, intrusive, viscous, fissure

Teaching and learning

Engage

- **Pairs to fours** Ask the students to **brainstorm** what they already know about (igneous) rocks. [O1]

Challenge and develop

- **Think, pair, share** Ask the students why the oldest rocks on the planet are igneous rocks (refer back to Topic 4.2). Take feedback from the class and discuss. [O1]
- **Pair talk** The students read the 'What are igneous rocks?' section in the Student Book and **discuss** the answers to questions 1 and 2. Take feedback from the class and agree on answers. [O1]
- Melt some wax crayons into liquid 'magma' – see Technician's notes 3.4.10 – and let the wax cool to form an 'igneous rock'. The students must wear safety glasses while **observing** this demonstration. Model cooling at different rates by letting some wax cool at room temperature and dropping some into water to speed up cooling. Relate the odd shapes of the water-cooled wax to the bubbles in extrusive volcanic rocks. Ask the students to **evaluate** the model. Take feedback and discuss their ideas. [O1&3]

Explain

- **Pair work** Working in pairs, the students complete task 1 of Worksheet 3.4.10. [O1]
- Still in their pairs, the students work on the practical activity to **model** and **investigate** magma flow, following the instructions on Practical sheet 3.4.10a. Relate the rate of flow of the hot magma to the rate of cooling and crystal size – e.g. high viscosity lava flows slowly, cools slowly and will have larger crystals than low viscosity lava. [O2]

- Ask the students to read the 'Looking at magma' section in the Student Book and **discuss** the answers to questions 3 and 4 in pairs. Take feedback and agree on answers. [O2]

  Higher-attaining students could work in pairs to complete a second practical activity, **investigating** how the rate of cooling affects crystal size, using Practical sheet 3.4.10b. [O3]

- Ask the students to read the final part of the Student Book and answer questions 5 and 6. [O3]

Consolidate and apply

- **Fours** Ask the students to complete task 2 of Worksheet 3.4.10. They can then **peer-assess** one another's posters. [O2]
- **Pair talk/Pairs to fours** The student pairs **devise questions** (along with answers) on the topic that start with 'What', 'Why', 'When', 'Where' and 'How'. Each pair asks another pair their questions. Take feedback from the class. What did they find easy? What did they find difficult? [O1, 2&3]
- **Fours** Ask the students to complete task 3 of Worksheet 3.4.10. If they struggle with part b, give them a clue – 'friction'. [O2&3]

Extend

- Ask students able to progress further to **research** igneous intrusions and **design** an information leaflet or web page. [O1, 2&3]

Plenary suggestions

Five key facts Ask the students to come up with five key facts that they have learned in this lesson. Take feedback and discuss. [O1, 2&3]

Answers to Student Book questions

1. for example: pumice; obsidian; basalt; granite; pegmatite; gabbro
2. they have no layers; usually contain crystals; rarely react with acid; do not contain fossils
3. acidic magma is viscous, so does not travel far; alkaline magma is thin, runny, and can travel long distances
4. Viscous magma tends to produce steep, conical volcanoes. Non-viscous magma tends to give shallow, sloping-sided volcanoes.
5. the faster the cooling, the smaller the crystal size
6. Intrusive rocks generally cool more slowly than extrusive rocks. However, magma in fissures can cool quickly to form rocks with small crystals; deep lava can cool very slowly to form extrusive rocks with large crystals.

Answers to Worksheet 3.4.10

1. a) molten rock cools and hardens b) *intrusive* igneous rocks formed below the ground; *extrusive* igneous rocks formed above the ground
2. a) *acidic* magma is viscous and so does not travel far; it will cool more slowly and have larger crystals; *alkaline* magma is thin and runny; it can travel long distances; it will cool more quickly and have smaller crystals
3. a) i) In trench A cooling is slower and forms larger crystals than in trench B, because less of the lava is exposed to the air. Lava at W will cool faster and form smaller crystals than lava at X, because it is closer to the surface. Lava at Y will cool faster and form smaller crystals than lava at Z, because it is closer to the surface. ii) Lava in deep trenches cools more slowly and forms bigger crystals than lava in wide trenches. Lava at the surface cools faster and forms smaller crystals than lava at the centre (or bottom) of the trench. iii) larger crystals in rocks formed from lava at X; it will cool more slowly

   b) basaltic lava will flow fastest; thicker liquids have more friction and so flow more slowly than thinner liquids; lava that contains less silica is less viscous and so flows faster; also, its temperature is higher; the hotter the lava is, the more quickly it will flow

Answers to Practical sheet 3.4.10a

1. (students' own graphs)
2. viscosity 0% < 20% < 40% < 60% < 80% 3. Low-viscosity materials flow more quickly. 4. repeat each concentration, e.g. 3 times 5. put a small coloured circle, e.g. from a punched hole, on the liquid to act as a marker

Answers to Practical sheet 3.4.10b

1. The quicker the rate of cooling, the smaller the crystal size.
2. granite cooled more slowly than basalt; granite has the larger crystals
3. repeat the experiment a number of times and measure the crystal size

4.11 Studying sedimentary rocks

Lesson overview

Learning objectives

- Describe how sedimentary rocks are formed.
- Explain how fossils give clues about the past.
- Explain the properties of sedimentary rocks.

Learning outcomes

- Describe sedimentation in layers; name and describe three examples of sedimentary rocks. [O1]
- Describe what a fossil is and explain why these are found in sedimentary rocks. [O2]
- Explain how rocks turn into sediments. [O3]

Skills development

- Thinking scientifically: ask questions
- Working scientifically: develop explanations
- Learner development: collaborate effectively

Resources needed smooth rounded pebbles or close-up pictures of these; samples of limestone, sandstone and mudstone; materials and equipment as detailed in the Technician's notes; Worksheet 3.4.11; Practical sheet 3.4.11a; Practical sheet 3.4.11b; Technician's notes 3.4.11

Digital resources Quick starter; Video; Slideshow: Clues from fossils - What do fossils tell us?; Interactive activity: Reorder the sentences to describe the freeze-thaw process

Key vocabulary weathered, deposition, sedimentary, fossil, freeze–thaw

Teaching and learning

Engage

- **Brainstorm in groups** Ask the students what they know about fossils. Take feedback, discuss their ideas and write them on the board for later. [O2]
- **Think, pair, share** Show the students some rounded pebbles. Ask them to **suggest** three ways that the pebbles could have been made so smooth. Take feedback, discuss their ideas and write them on the board for later. [O3]

Challenge and develop

- **Group work** Ask the students to work in groups to **carry out** a practical activity to **model** making limestone using Practical sheet 3.4.11a. They should see a precipitate of calcium carbonate, which should settle. When the clay suspension is added, a layer of clay sediment should form on top of the 'limestone'. The students should then answer the questions on the Practical sheet. [O1]
- Give the students samples of limestone, sandstone and mudstone – allow the students to **examine** them using a hand lens. Ask them to **describe** what the lumps of rock consist of. Use questioning to help them to **deduce** that they are made of particles stuck together. [O1]
- Students read the 'Rocks in layers' section in the Student Book, and **discuss** and answer questions 1 and 2. Take feedback. [O1]

Explain

- Set up a demonstration of sedimentation – see Technician's notes 3.4.11. This needs to be done outside. Ask the students to **watch** as you switch on the water at the top end, using a spray head such as on a watering can. Ask the students to answer the questions on Practical sheet 3.4.11b. Establish that smallest particles are transported most easily and are the last to be deposited. Relate this to deposition of sediments as rivers enter seas – pebbles are deposited first and finer sediments last. Ask the students why different layers of sediment may form. [O1&3]

- **Pair work** Ask the students to complete task 1 of Worksheet 3.4.11 in pairs. They **sequence** the stages of sedimentary rock formation and **describe** how limestone and sandstone are made. [O1]
- **Twos to fours** In pairs, the students read the 'Looking at fossils' section in the Student Book and answer questions 3 and 4. They can then **devise** 'What', 'Why', 'Where', 'When' and 'How' questions about fossils, along with answers, and ask another pair their questions. [O2]

Consolidate and apply

- Ask the students, working individually, to **describe** what a fossil is and **explain** why they are found in sedimentary rocks. They should then **peer-assess** the work of a partner using task 2 of Worksheet 3.4.11. [O2]
- **Think, pair, share** The students read the 'Breaking rocks' section in the Student Book and answer questions 5 and 6. [O3]

Extend

- Ask students able to progress further to work in pairs to **research** information and **produce a leaflet** on limestone and quarrying in Yorkshire. [O1&3]

Plenary suggestions

Cartoon strip Ask the students to work in pairs and draw a cartoon strip **explaining** one of the processes learned about in the lesson – sedimentation, the formation of fossils or the weathering of rock using task 3 of Worksheet 3.4.11). [O1, 2&3]

Answers to Student Book questions

1. Tiny rock pieces are broken off by wind or water (also by freeze-thaw weathering, tree roots, etc.) and are deposited as sediments in water.
2. Any three from: *sandstone* is made from sand particles; *limestone* is made from remains of tiny shells and micro-skeletons of marine organisms deposited on the seabed; *shale/ mudstone* is made from small silt and clay-sized particles; *conglomerates* are made from rounded pebbles cemented together.
3. In the sea; contains shells and skeletons of sea creatures.
4. Animals or plants die, become covered in sediments, and the sediments turn to rock; some organisms are preserved in amber, glaciers, etc.
5. They bump into each other on the riverbed/seabed as they are carried in water; the sharp edges are knocked off.
6. *Deposition*: rocks and pebbles fall out of slow-moving water to the riverbed/seabed. *Compaction*: water is squeezed from lower layers by the mass and pressure from upper layers. *Cementation*: water seeps through the sediments and minerals in it can crystallise between the rock particles and cement them together.

Answers to Worksheet 3.4.11

1. a) B; A; H; G; F; C; E; D

   b) i) Sandstone is made from sand particles. ii) Limestone is made from remains of tiny shells and micro-skeletons of marine organisms deposited on the seabed.
2. a) A fossil is the remains of a dead plant or animal, preserved in rock.

   b) Fossils are found in sedimentary rock because they become covered/trapped in layers of sediments, which over time become sedimentary rock.
3. Idea and descriptions of physical, chemical and biological weathering, causing rock pieces to be washed into rivers etc.; then rounding off of pebbles that get smaller and smaller to form sediments.

Answers to Practical sheet 3.4.11a

1. calcium carbonate
2. build their shells
3. They sink into the sediment and form fossils in the rock.
4 It is deposited at the bottom of the sea.
5. It must be weathered and eroded.

4.12 Using metamorphic rocks

Lesson overview

Learning objectives

- Name some examples of metamorphic rocks.
- Describe how metamorphic rocks are formed.
- Explain why metamorphic rocks are suited to their uses.

Learning outcomes

- Describe how metamorphic rocks are formed and give examples. [O1]
- Describe the properties of metamorphic rocks and link this to the way that they are formed. [O2]
- Explain why the properties of some metamorphic rocks make them suitable for different uses. [O3]

Skills development

- Thinking scientifically: interpret evidence
- Working scientifically: develop explanations
- Learner development: collaborate effectively

Resources needed materials and equipment as detailed in the Technician's notes; Worksheet 3.4.12; Practical sheet 3.4.12; Technician's notes 3.4.12

Digital resources Quick starter; Interactive activity: Drag the statements to the correct group - metamorphic or rocks of other types; Interactive activity: Match the materials before and after metamorphic change

Common misconceptions Each type of rock can only be metamorphosed into one particular type of metamorphic rock. A metamorphic rock cannot be metamorphosed. All fossils are always destroyed during metamorphosis of rocks.

Key vocabulary metamorphic, metamorphose, recrystallise

Teaching and learning

Engage

- **Brainstorm** Explain that existing rocks can be changed into metamorphic rocks. Ask the students 'What could cause metamorphic rocks to form?'. Discuss ideas from the class. [O1]
- **Think, pair, share** Give the students some Plasticine (or other modelling clay) – this is to represent some rock. They make a ball – ask them 'What causes the Plasticine to change shape?' Elicit that pressure (from their muscles) and heat (from hands and friction) softens/reshapes the Plasticine. Ask the students 'Where do heat and pressure come from in the Earth?' Stress that with enough pressure and heat, rock like clay can form a new type of rock called 'metamorphic rock'. Sometimes molten rock forces itself into cracks in other types of rock. The very hot magma (700 to 1400 °C) metamorphoses the surrounding rock. [O1]

Challenge and develop

- The students **observe** a demonstration that models metamorphosis. Put an egg white in a transparent dish. Then add 50 cm^3 of water to a beaker, add two or three spatulas of sodium chloride (to increase the boiling point) and heat to boiling over a Bunsen burner. Take the beaker off the tripod and put it in the centre of the egg white – leave it for 10 minutes. Ask the students to **observe** the egg white. Ask 'What has happened? 'Can the change be reversed by cooling it again?'. Discuss how this models the metamorphosis of rocks. (Uncooked egg white is like molten magma – it changes when exposed to heat, and the change cannot be reversed.) [O1]
- Demonstrate a simulation of the formation of slate sheets. (Slate is formed when pressure is exerted on mudstone or shale.) Pour some used matches or short pieces of spaghetti randomly over the bench, so that

they lie in all directions. These represent the microscopic clay minerals in mudstone or shale. Take two 30 cm rulers and put them on either side of the matchsticks/ spaghetti. Push the rulers together forcing the 'clay minerals' to line up parallel to the moving rulers. This models clay minerals being forced to line up at right angles to the direction of applied forces – for example forces from two approaching plates. [O1]

- **Think, pair, share** The students read the 'Making metamorphic rocks' section in the Student Book. They **discuss** and **agree on** answers to questions 1 and 2. [O1]

Explain

- Student pairs **discuss** task 1 of Worksheet 3.4.12 and **agree on** the correct sequence. They **draw annotated diagrams** to illustrate each stage of the process and **peer-assess** another pair's work. [O1]
- Read and discuss the 'Metamorphic changes' section in the Student Book as a class. Ensure that the students understand that rocks recrystallise without melting, that the changes are permanent and that different levels of heat and pressure result in different types of minerals. [O2]
- **Group work** The students **examine** labelled samples of sandstone, limestone, mudstone, quartzite, marble and slate using a hand lens. They **classify** these as sedimentary and metamorphic rocks and **explain** how they are different (question 3 in the Student Book). Take feedback from different groups and agree on the answers. [O2]

Consolidate and apply

- **Pair work** Ask the students to complete task 2 of Worksheet 3.4.12. They need to **explain** what information scientists can deduce from crystals in rocks, and **identify** the pressure acting on trilobite fossils. [O2]
- Continuing in their pairs, the students **follow the instructions** on Practical sheet 3.4.12 to **investigate** the effect of pressure on fossils in rocks. Squeezing top to bottom causes small, fat fossils; squeezing side to side causes tall, thin fossils; squeezing one side up and one down causes tall, twisted fossils. Link this to scientists using fossils to deduce movements of the Earth's crust. [O1&2]
- **Pair talk/Pairs to fours** The students read the 'Metamorphic rocks in detail' section in the Student Book, **discussing** and **agreeing on** answers to questions 5 and 6. [O3]

Extend

- Ask students able to progress further to complete task 3 of Worksheet 3.4.12. Each student **researches** a metamorphic rock. In pairs or groups, they **combine** the information they have each collected and **present** it to other students, who should **peer-assess** the work. [O3]

Plenary suggestions

The big ideas The students write down three ideas they learned during the lesson. They **share** their facts in groups and **compile** a prioritised master list of facts. Take feedback and find out whether groups agreed or not. [O1, 2&3]

Answers to Student Book questions

1. by heat and/or pressure acting on existing rocks 2. sedimentary
3. metamorphic rocks are harder; some are shiny; original features (e.g. layers and fossils) are destroyed or altered
4. the heat melts them and the pressure squashes them or destroys them
5. Different rocks are formed from the same original rock, depending on the amount of heat and pressure applied.
6. the layers in sedimentary rocks, fossils, and the clues they give us to the origin of the rock have usually been destroyed

Answers to Worksheet 3.4.12

1. a) C; D; B; A
2. a) Different minerals form at different temperatures and pressures. The new minerals can be used to estimate the temperature, depth and pressure at which the original rock was metamorphosed.
   b) they might have been distorted or destroyed c) horizontally squeezed by 50% d) vertically squeezed by 66%
   e) it suggests that the rocks also underwent the same percentage distortion
3. *Quartzite*, *marble* and *slate*: see the Student Book; all formed at high temperature and pressure; *schist*: from slate exposed to further heat and pressure; used as decorative rock in gardens; *gneiss*: from slate; extreme heat and pressure; used in building; *serpentine*: from dunite/peridotite; in mountains and near seabed; by slightly raised temperature; used as a carving stone and in building

4.13 Understanding the rock cycle

Lesson overview

Learning objectives

- Describe the rock cycle.
- Explain how rocks can change from one type to another.

Learning outcomes

- Describe simply how rocks can be changed from one type to another. [O1]
- Explain the processes involved in the rock cycle using scientific language. [O2]
- Explain how the properties of each rock type link to the processes in the rock cycle. [O3]

Skills development

- Thinking scientifically: ask questions
- Working scientifically: develop explanations
- Learner development: communicate effectively

Resources needed materials and equipment as detailed in the Technician's notes; large sheets of paper; Worksheet 3.4.13a; Worksheet 3.4.13b; Practical sheet 3.4.13; Technician's notes 3.4.13

Digital resources Quick starter; Interactive activity: Place the processes of the rock cycle and the types of rock in the correct sequence, starting with 'Weathering and erosion'; Slideshow: Layers and folds - A look at how rocks are weathered; Hangman: Key vocabulary game

Common misconceptions Mountains are unchanging.

Key vocabulary rock cycle, uplift, upfold, anticline, syncline

Teaching and learning

Engage

- **Think, pair, share** Challenge the students to **link** the different rock types together in a mind map and name the processes involved. [O1]
- **Pairs to fours** Alternatively student pairs could **list** points they already know about the rock cycle. Fours then **discuss** it further and come up with a list of points. One student **reports** back – start to develop the rock cycle on the board as the students report back. [O1]

  Higher-attaining students may be able to attempt this independently.

Challenge and develop

- **Pair talk** Students read the 'The rock cycle' section in the Student Book and **study** Figure 3.4.13a. They then **discuss** the answers to questions 1 and 2. Take feedback and **agree on** answers. [O1]
- Arrange the class for a **team memory game** about the rock cycle. The students' instructions are given on page 1 of Worksheet 3.4.13a. The diagram of the rock cycle, to be shown to students for only brief periods, is given on page 2 of the worksheet. Encourage the students to **ask questions** in their teams, to clarify their understanding. [O1]
- The students can then complete task 1 of Worksheet 3.4.13b. [O1]

Explain

- **Pair work** Students work in pairs to complete Practical activity 3.4.13 to **model** how rocks fold and fault. Most will probably need to be shown how to make reasonably even horizontal layers.

  Some groups of students may find the care and time needed is too much, in which case you can demonstrate the activity.

- Ask the students to **evaluate** the model from the practical activity. They should think about the spatial scale, the time scale and the lack of heat. [O2]
- The students read the 'Folding rocks' section in the Student Book and **discuss** the answers to questions 3 and 4 in their pairs. Take feedback from the class and **agree on** answers. [O2]
- **Group work** Ask the students to **look carefully** at samples of sedimentary rock, igneous rock and metamorphic rock and **compare and contrast** their appearance and structure. They should think about colour, whether they have crystals or not, whether they are shiny or not, whether they have observable particles or not, and how the particles are arranged. Ask them to **draw** each rock in detail and **label** their drawings fully. [O1, 2&3]
- Ask the students to read the 'Rocks on the move' section in the Student Book and answer questions 5 and 6. [O3]

Consolidate and apply

- The students carry out task 2 of Worksheet 3.4.13b to **explain** how jagged, weathered pieces of rock become smooth round pebbles. Take feedback. [O2]

  Lower-attaining students may need a hint here: wind and water.

- Ask the students to **imagine** that they are particles in molten magma – what happens to them? The students can complete a piece of **extended creative writing** to tell the story. [O1, 2&3]

Extend

- Ask the students to complete task 3 of Worksheet 3.4.13b, **linking** the properties of the different types of rocks to the processes that formed them. [O3]

Plenary suggestions

Mind map The students revisit their mind map from the *Engage* activity and add to it to show their learning about the structure of the Earth, different types of rocks and the rock cycle. [O1, 2&3]

Answers to Student Book questions

1. sedimentary, metamorphic and igneous rocks can be changed to sedimentary rocks by weathering, erosion and sedimentation; sedimentary rocks can be changed to metamorphic rocks by heat and pressure; sedimentary and metamorphic rocks can be changed to igneous rocks by heat (melting and then cooling to form the new rock)
2. *weathering* means cracking rocks and breaking pieces off; *erosion* means breaking down rocks further into particles, which are then transported, e.g. to the sea
3. a) wind and water break up larger rock pieces into much smaller rock pieces; this allows them to be transported, again by wind and water

   b) small rock particles are deposited on a riverbed/seabed as sediments when the water (current) slows down. They are buried by new sediments forming on top of them; the new sediments compact the lower layers and squeeze the water out to form sedimentary rock

   c) sedimentary and other types of rock metamorphose with heat and pressure to form new metamorphic rock. Heat can also melt rocks under the crust to form magma; magma cools to form igneous rocks below and above the Earth's surface. Pressure can also cause rock folding
4. these can be formed when rocks fold because of surrounding rock movements; *anticlines*: strata are pushed up in a dome shape; *synclines*: strata are forced down in a bowl shape
5. by uplift, which can occur due to the constant movement of the Earth's crust – for example during an earthquake
6. *quickly*: uplift; volcanic eruption; folding; faulting; *slowly*: weathering; erosion; sedimentation; metamorphosis, melting

Answers to Worksheet 3.4.13b

1. a) 1 with F; 2 with G; 3 with D; 4 with E; 5 with C; 6 with A; 7 with B b) See Figure 3.4.13a in the Student Book.
2. they should include the ideas that weathered pieces of rock knock against each other and are further weathered; protruding edges are knocked off; some pieces of rock fall into streams and are eroded further as they travel in the water, and as they knock into each other gradually they become smooth
3. *sedimentary*: softer, layered rocks because the particles are just compacted and cemented together; this makes them rough; the spaces between compacted particles make the rocks porous; fossils are usually found in sedimentary rocks, because there is no heat/pressure involved in the rocks' formation; *igneous*: interlocking crystals that formed as they solidified; they are hard with no layers; many have no gaps and are not porous; their properties depend on their composition and the temperature and pressure; *metamorphic*: hard because the minerals are lined up or arranged in uneven layers; their properties depend on the heat and pressure the rocks have been subjected to; any fossils found are often distorted

4.14 Checking students' progress

The 'Checking your progress' section in the Student Book indicates the key ideas developed in this chapter and shows how students progress to more complex levels. It is provided to support students in:

- identifying those ideas
- developing a sense of their current level of understanding
- developing a sense of what the next steps in their learning are.

It is designed either to be used at the end of a chapter to support an overall view of the progress, or alternatively during the teaching of the chapter. Students can self-assess or peer-assess using this as a basis.

It would be helpful if students can be encouraged to provide evidence from their understanding or their notes to support their judgements. In some cases it may be useful to explore the difference in the descriptors for a particular idea so that students can see what makes for a 'higher outcome'.

It may be useful with some descriptors to provide examples from the specific work done, such as an experiment undertaken or an explanation developed and recorded. If marking and feedback uses similar ideas and phrases, this will enable students to relate specific marking to a more general sense of progress.

To make good progress in understanding science students need to focus on these ideas and skills:

Students who are making modest progress will be able to:	Students who are making good progress will be able to:	Students who are making excellent progress will be able to:
Name the main gases in the atmosphere and describe the composition of the atmosphere.	Describe how the gases in the atmosphere, and their relative proportions, have changed over time.	Explain how plants and then animals have changed the atmosphere over time using different processes.
Name and describe some examples of human activities that cause damage to the atmosphere.	Name and describe the environmental effects caused by some human activities and suggest how they can be reduced.	Name some technological advances that reduce the effects of human activities on the atmosphere and analyse their impact.
Identify natural resources that the Earth provides.	Explain how human activities limit or damage these resources – for example through agricultural practices.	Analyse the advantages of a development against the disadvantages to the environment and make evidenced judgements about whether the development should be allowed.
Describe what global warming is and identify the effects of global warming on the Earth.	Explain how global warming affects living organisms and identify that scientists across the world have different opinions about global warming and its effects.	Evaluate the arguments for and against human activities enhancing the global warming effect.
Describe the carbon cycle as natural recycling and name the different ways that carbon enters and leaves the atmosphere.	Explain how each stage of the carbon cycle affects the amount of carbon in the atmosphere.	Explain what is meant by 'carbon footprint' and analyse the carbon footprint of different activities.
Describe the benefits of recycling and name materials that can be recycled.	Explain the advantages and disadvantages of some recycling schemes.	Compare the efficiency of different recycling schemes to extracting the natural resource from the Earth.
Name the layers that make up the Earth and identify that the Earth's surface is made of plates that move constantly.	Describe the characteristics of each layer of the Earth and recall that tectonic plates move very slowly.	Explain that earthquakes, volcanic eruptions and the formation of mountains can happen where tectonic plates meet; explain how volcanic activity changes the surface of the Earth.
Describe how igneous, sedimentary and metamorphic rocks are formed; give examples and describe how they can change from one type to another.	Describe the features and properties of these rocks, including crystals in igneous rocks, recrystallisation in metamorphic rocks and layers (burying fossils) in sedimentary rocks.	Explain the processes involved in the rock cycle and link these to how the rocks are formed.

4.15 Answers to Student Book Questions

This table provides answers to the Questions section at the end of Chapter 4 of the Student Book. It also shows how different questions assess attainment in terms of the focus and style of a question as well as the context. Question level analysis can indicate students' proficiency in approaching different aspects of scientific understanding and different types of answer.

Q	Answer		Focus			Style			Context			
		Marks available	Knowledge & understanding	Application	Evaluation of evidence	Objective test question	Short written answer	Longer written answer	The atmosphere	Earth's resources and recycling	Structure of the Earth	The rock cycle
1	c	1	x			x			x			
2	d	1	x			x					x	
3	b	1	x			x					x	
4	a	1	x			x			x			
5	• more UV light will reach the Earth's surface • higher risk of skin cancer	1 1	x x				x x		x x			
6	• separate solid parts of the Earth's lithosphere • that move about very slowly	1 1	x x				x x				x x	
7	any **two** of: • overfishing; depletes fish stocks/destroys marine food chains/ webs • intensive farming; depletes soil of minerals, destroys habitats, overgrazing destroys grass etc. • mining/ quarrying; scars landscape and destroys habitats, pollutes water systems etc. • landfill; pollutes water and soil, destroys habitats etc. • use of fertilisers and other chemicals; pollutes water systems, bioaccumulation etc. • any other correct example, with impact	4	x					x		x		
8	a	1		x		x					x	
9	c	1		x		x					x	
10	• water between rock particles is gradually squeezed out • by the increasing weight of new deposits on top	1 1		x x			x x					x x

Q	Answer		Focus			Style			Context			
		Marks available	Knowledge & understanding	Application	Evaluation of evidence	Objective test question	Short written answer	Longer written answer	The atmosphere	Earth's resources and recycling	Structure of the Earth	The rock cycle
11	• it is an igneous rock	1		x				x				x
	• that formed from molten rock/ magma/ lava	1		x				x				x
	• this cooled quickly	1		x				x				x
	• so it only had time to form the small crystals observed	1		x				x				x
12	D	1		x		x				x		
13	A	1		x		x				x		
14	• a dead animal on the seabed was covered by sediments	1		x			x					x
	• soft tissues broke down leaving the hard parts in the (sedimentary) rock	1		x			x					x
15	• weathering/erosion	1			x		x					x
	• by wind/ rain	1			x		x					x
16	• the lower the pH, the fewer lakes there are with high fish populations	1			x			x	x			
	• and the more lakes there are with no fish (or converse).	1			x			x	x			
	• acid rain formed (by burning fossil fuels) makes the water acidic	1			x			x	x			
	• and acidic water kills the fish/kills their food/fewer fish are adapted to low pH	1			x			x	x			
	Total possible:	**30**	**12**	**12**	**6**	**8**	**10**	**12**	**8**	**6**	**6**	**10**

5.1 Introduction

When and how to use these pages

The Introduction in the Student Book indicates some of the ideas and skills in this topic area that students will already have met from KS2 or from previous KS3 work, and provides an indication of what they will be studying in this chapter. *Ideas you have met before* is not intended to comprehensively summarise of all the prior ideas, but rather to point out a few of the key ones and to support the view that scientific understanding is progressive. Even though students might be meeting contexts that are new to them, they can often use existing ideas to start to make sense of them.

In this chapter you will find out indicates some of the new ideas that the chapter will introduce. Again, it isn't a detailed summary of content or even an index page. Its purpose is more to act as a 'trailer' and generate some interest.

The outcomes, then, will be recognition of prior learning that can be built on, and interest in finding out more.

There are a number of ways this can be used. You might, for example:

- Use *Ideas you have met before* as the basis for a revision lesson as you start the first new topic.
- Use *Ideas you have met before* as the centre of spider diagrams, to which students can add examples, experiments they might have done previously or what they found interesting.
- Make a note of any unfamiliar/difficult terms and return to these in the relevant lessons.
- Use *In this chapter you will find out* to ask students questions such as:
  - Why is this important?
  - How could it be used?
  - What might we be doing in this topic?

Overview of the chapter

In this chapter, the students will develop their ideas about forces and motion, and move to more qualitative approaches. They will collect data and use it to plot and interpret distance–time graphs for single objects and for objects moving in relation to one another.

The students will explore the equilibrium of forces in static and dynamic situations and what happens when the forces are not in equilibrium. They will also learn how forces acting in the same dimension can be added together to increase the effect, and how one force can be subtracted from another leading to a reduced effect.

The chapter returns to the concept of gravitational fields and quantifies the strengths of fields in relation to the masses of planets and the distances from them. The concept of gravitational field will be applied to the arrangement and movement of objects in the Solar System. The students will learn about the effects of the Earth's motion in relation to the Sun, including the implications of the Earth's tilted axis. They will describe some characteristics of our Sun in comparison to other stars. They will consider the magnitude of space, the measurement of distances in light years and the limitations of measuring in kilometres.

Throughout the chapter, the students are challenged to deepen their understanding of both familiar and unfamiliar ideas and apply them to a range of situations.

Obstacles to learning

The students may need extra guidance with the following concepts:

- **Forces** Many students think that there are no forces acting when an object is stationary, and that for an object to move there must be a force acting on it to keep it moving. In vehicle collisions, it can appear that a passenger suddenly flies forward as opposed to carrying on at the same speed and direction (Newton's First Law). For orbiting objects, students may think that if the force keeping the object in orbit is removed, the object will fly outwards, rather than at a tangent. Gravitational field strength is what determines how fast an object falls – not its mass or weight.
- **Seasons** On Earth, the seasons are sometimes thought to be caused by the Earth being further away from the Sun at certain points during its elliptical orbit. In fact, the tilt of the Earth on its axis is the biggest factor.

- **Stars** Some students think that stars are balls of burning gas because they may not be familiar with nuclear fusion, so will not appreciate the difference between this and burning. Planets and stars may appear fundamentally the same to some students because both glow in the night sky.
- **Galaxies** Interstellar distances are difficult to envisage and are many orders of magnitude greater than distances within the Solar System. The unit of distance – the light year – is sometimes thought of as a measure of time.

	Topic title:	**Overarching objectives:**
2	Describing journeys with distance–time graphs	The representation of a journey on a distance–time graph
3	Exploring journeys on distance–time graphs	The representation of a journey on a distance–time graph Speed and the quantitative relationship between average speed, distance and time (speed = distance ÷ time)
4	Understanding relative motion	Relative motion: trains and cars passing one another
5	Analysing equilibrium	Opposing forces and equilibrium: weight held by stretched spring or supported on a compressed surface Using force arrows in diagrams, adding forces in one dimension, balanced and unbalanced forces
6	Exploring motion and equilibrium	Opposing forces and equilibrium: weight held by stretched spring or supported on a compressed surface Using force arrows in diagrams, adding forces in one dimension, balanced and unbalanced forces Forces being needed to cause objects to stop or start moving, or to change their speed or direction of motion (qualitative only)
8	Understanding gravitational fields	Gravity force, weight = mass × gravitational field strength (*g*), on Earth *g* =10 N/kg, different on other planets and stars
9	Applying ideas about gravitational fields	
10	Looking at motion in the Solar System	Gravity forces between Earth and Moon, and between Earth and Sun (qualitative only)
11	Describing stars and galaxies	Our Sun as a star, other stars in our galaxy, other galaxies
12	Explaining the effects of the Earth's motion	The seasons and the Earth's tilt, day length at different times of year, in different hemispheres
13	Measuring distances in the Universe	The light year as a unit of astronomical distance

5.2 Describing journeys with distance–time graphs

Lesson overview

Learning objectives

- Gather relevant data to describe a journey.
- Use the conventions of a distance–time graph.
- Display the data on a distance–time graph.

Learning outcomes

- Gather evidence to construct a distance–time graph. [O1]
- Construct a graph to represent a journey. [O2]
- Construct a graph to represent a more complex journey. [O3]

Skills development

- Thinking scientifically: analyse data
- Working scientifically: present evidence
- Learner development: ask questions

Resources needed dynamics trolley; ramp; wind-up toy cars; digital timers; metre rulers; graph paper; calculators; Worksheet 3.5.2; Practical sheet 3.5.2; Technician's notes 3.5.2

Digital resources Quick starter; Slideshow: Time-lapse photography; Interactive activity - Match the sentences about distance–time graphs for a car journey

Common misconceptions A straight line sloping upwards on a distance–time graph indicates that speed is increasing.

Key vocabulary distance–time graph, accelerate, stationary

Teaching and learning

Engage

- Demonstrate a simple journey by rolling a dynamics trolley down a ramp, across the floor and allowing it to come to rest when it reaches a wall or due to friction. The students **describe** the journey in words. [O1]
- Ask some students to **share** their descriptions with the class. Highlight the good points, such as students recognising how speed and distance from the start position changed during the journey. [O1]

  Higher-attaining students can be encouraged to **use scientific terms** such as 'accelerate' and 'constant speed' in their descriptions, where appropriate.

Challenge and develop

- Explain to the students that they will learn to **translate** their journey descriptions into distance–time graphs. As an introduction to this, they **plot** a distance–time graph using the data in task 1 of Worksheet 3.5.2, which represents travel at a constant speed. [O2]
- Using Figures 3.5.2b and 3.5.2c in the Student Book, show the students some graphs for different simple movements. [O1&2]
- They **match** graphs to descriptions of different journeys in task 2 of Worksheet 3.5.2. [O1&2]

Explain

- **Pair work** Using Practical sheet 3.5.2, the students work in pairs to **observe** the motion of a mechanical toy and **write a description** of its journey. They should **present** their evidence by **sketching** a distance–time graph. They should then annotate the graph to **explain** what the toy is doing at significant parts of the graph – for example 'The toy is released here and moves away', 'The line is straight and this shows that the toy is moving at constant speed'. Ask the students to think about the accuracy of the data they have presented, and how this might be improved. [O1, 2&3]

Higher-attaining students may be able to **measure** the distance travelled and the time so that they can add a scale to their graph.

Consolidate and apply

- The students **analyse data** by answering questions 5–9 in the Student Book. [O2&3]
- They use quantitative data to **plot** a distance–time graph accurately in task 3 of Worksheet 3.5.2. [O2&3]

Extend

- Ask students able to progress further to **suggest** what a distance–time graph would look like if, after travelling some distance, someone turned around and went back to where they started. As time passes, the distance from the start position *decreases*, so the line slopes down from left to right until it meets the *x*-axis. [O3]

  Note: At this stage they are not asked to consider displacement, so the line does not return to the origin.

Plenary suggestions

Whole class discussion Select a group to **sketch** a distance–time graph on the board so that the whole class can **share** it. Members of the group ask questions to other students to test their **interpretation** and **understanding** of distance–time graphs. [O1, 2&3]

Answers to Student Book questions

1. m/s; metres per second
2. 50 m
3. 5 m/s
4. The second line should be half as steep, indicating that half the distance is covered in the same time; 25 m in 10 s.
5. a) The object is not moving.
   b) The object is moving at a steady speed.
   c) The object is accelerating.
6. from 45 s onward
7. The line is steeper; more distance is covered during each minute, so the bus must be moving faster.
8. Section 1 is slightly steeper than section 3; in section 3 the student was walking with a friend, so they may have been talking and walking more slowly.
9. Numerical scales are not needed but students who have progressed well may be able to include them. The labels should show clearly that they appreciate the main features of a distance–time graph: upward slope indicates movement; horizontal line indicates no movement; steep slope indicates faster speed than a less steep slope; upward curve indicates acceleration.

Answers to Worksheet 3.5.2

1. a) Straight line through the origin; *x*-axis = time; *y*-axis = distance; axes labelled; scales and units included. b) iii
2. a) vi b) iii c) ii d) v
3. a) Points plotted accurately; axes and the two curves labelled; scales and units included. Some students may consider that the changes in speed are not abrupt and will smooth out the transitions.
   b) i) between 10 minutes and 15 minutes ii) between 20 minutes and 30 minutes, and after 50 minutes when the journey has ended
   c) i) 90 minutes ii) 50 minutes
   d) Both achieved 20 km in 10 minutes; speed = distance travelled ÷ time taken = 20 ÷ 10 = 2 km/minute = 120 km/h.
   e) Car travels 80 km in 90 minutes; speed = 80 ÷ 90 = 0.89 km/minute = 53.3 km/h.
   Train travels 80 km in 50 minutes; speed = 80 ÷ 50 = 1.6 km/minute = 96 km/h.
   f) The car has to travel more slowly in town, stopping for traffic lights and in congestion. The train encounters fewer delays and only has to stop occasionally at stations.

Answers to Practical sheet 3.5.2

4, 5. Not accurate; it is difficult to judge the changes in speed; the graph shows information for motion in a straight line – the path of the car may curve.

6. Ensure the motion is along a straight track; fix a card to the car and use light gates and a datalogger/use time-lapse photography.

5.3 Exploring journeys on distance–time graphs

Lesson overview

Learning objectives

- Interpret distance–time graphs to learn about the journeys represented.
- Relate distance–time graphs to different situations and describe what they show.

Learning outcomes

- Explain some of the features of a distance–time graph. [O1]
- Explain what is represented on a distance–time graph. [O2]
- Explain what is represented on a more complex distance–time graph. [O3]

Skills development

- Thinking scientifically: evaluate risks
- Working scientifically: record evidence
- Learner development: ask questions

Resources needed large sheets of paper, e.g. flip chart paper; marker pens; mini-whiteboards; equipment and materials as detailed in the Technician's notes; Worksheet 3.5.3; Practical sheet 3.5.3; Technician's notes 3.5.3

Digital resources Quick starter; Video; Interactive activity: Drag the statements into the correct group - those which describe acceleration, and those which do not

Common misconceptions A straight line sloping upwards on a distance–time graph indicates that speed is increasing.

Key vocabulary distance–time graph, time-lapse sequence, accelerate

Teaching and learning

Engage

- Show a video clip of a sporting activity that involves running at different speeds and stopping, such as tennis or football. The students work in small groups and use a large sheet of paper to **sketch** a distance–time graph for the sporting activity. They should add captions to **describe** what the sportsperson is doing in each part of the graph. [O1, 2&3]
- The students **compare** their graphs to those of other groups, **ask questions** and make any modifications required. [O1, 2&3]

  Higher-attaining students should be encouraged to draw their graphs showing the sportsperson returning to their starting position.

Challenge and develop

- **Group work** If the technology is available, the students should work in groups to **collaborate effectively** to make time-lapse sequences to help them to **record evidence** and **analyse** the movement of different objects – like a ball or a trolley on a ramp. A clearly visible scale should be placed alongside each moving object so that its position can be observed at regular time intervals. The students use data from the time-lapse images to **plot** and **analyse** distance–time graphs. Before starting the practical activity, the students should **assess the risks**. (The main risks are from obstruction and tripping in the work area.) Practical sheet 3.5.3 gives instructions for this practical work and the analysis of the results. [O1, 2&3]

  If you do not have access to enough suitable cameras, this could be carried out as a teacher demonstration in which the students are involved in making important decisions and help to record the images.
- Show a video clip of an 800 m athletics race. Ask the students to **imagine** three athletes in the race – one who sets off running fast, gets tired and finishes in 2 minutes 15 seconds; one who runs at a slightly slower

pace but is able to sprint the last 50 m and finishes in 1 minute 58 seconds; and one who runs slower than the others for most of the race but then speeds up for the last 200 m and overtakes them to finish in 1 minute 48 seconds. [O2&3]

- **Pair talk** The students **discuss** the three athletes and **sketch** distance–time graphs to show the races. [O2&3]

Explain

- The students then **explain** what the graph(s) of the three athletes' races shows about their motion. [O2&3]

  Higher-attaining students could be asked to superimpose additional lines on the graph to show a 100 m sprinter in action and a marathon runner.

Consolidate and apply

- **Pair talk** The students **discuss** and **interpret** the distance–time graphs and answer the questions in the Student Book. [O1, 2&3]
- The students complete the tasks of Worksheet 3.5.3. [O1, 2&3]

Extend

- Ask students able to progress further to **calculate** acceleration from some of the distance–time graphs in the Student Book. For example, for Figure 3.5.3b they could calculate the speed for the first and second straight-line sections. They could then estimate the time span over which the change in speed took place. Acceleration = (initial speed – final speed) ÷ time taken to change speed. In this case, the answer will be negative because the car is slowing down. [O3]

Plenary suggestions

Mini-whiteboard quiz Select students to **ask questions** of the rest of the class about the lesson, which they answer on mini-whiteboards. Students will need to be primed to ask questions that require a short answer, such as a calculation or sketching a graph. [O1, 2&3]

Answers to Student Book questions

1. speed = distance travelled ÷ time taken 2. The line is not straight; the slope has changed.
3. Student reads the graph and uses the speed formula correctly; speed = 3 m/s in first section; 2 m/s in second section.
4. 25 minutes 5. 43.2 km/h 6. Graph should have linear scales: *x*-axis = time; *y*-axis = distance; straight, accurate line through (0, 0) and chosen point. 7. a) B b) C 8. m/s^2
9. a) The line is extended along the same slope, and then suddenly becomes horizontal. b) The straight line curves smoothly to become horizontal underneath the horizontal section of line A. c) The line becomes straight and shallow.

Answers to Worksheet 3.5.3

1. a) i) where the graph is steepest ii) where the graph curves upwards iii) where the upward curve of the graph gets shallower iv) where the graph is horizontal b) The force due to gravity/weight pulls it downwards so it gets faster; once it is level on the floor, friction slows it down. c) The spaces between the images increase until the ball leaves the ramp, and then reduce again. The last two images should be very close together, or even overlapping, because the ball is hardly moving as it comes to a stop.
2. a) 4 b) 460 m c) 210 m d) The ball travels fastest as it leaves the club. After that, the forces acting are air resistance and then friction with the ground, which slow the ball down until it stops. The curves show that the speed is changing; straight lines are where speed is not changing, e.g. the horizontal sections show the ball is stationary.
3. a) For a car travelling 1 km in 1 minute, speed = distance travelled ÷ time taken = 1 km/min = 60 km/h.

   b) Car A travels at a steady speed (60 mph) for the whole journey. Car B travels at a higher speed at first (approx. 1.5 miles in 1 minute = 90 mph) and then slows for the first camera (approx. 0.1 miles in 0.2 minutes = 30 mph). It then travels at its original speed again, before slowing for the second camera. Finally, car B speeds up to its original speed. Allow reasonable tolerance on reading data from the graph, but encourage precision. Also encourage them to convert speeds into 'per hour' rather than 'per minute'. c) Driver of car A would not be convicted. Driver of car B could be convicted even though it passes both cameras at 30 mph, because it takes approximately 0.8 minutes to travel 1 mile = 75 mph average speed between the two cameras.

Answers to Practical sheet 3.5.3

1. to 3. (students' own graphs) 4. A shorter time interval samples the data a higher number of times, so gives more information and makes the graph and analysis more accurate.
5. It allows frequent observations that could not be done by eye; the data is automatically recorded; an alternative would be light gates and a datalogger.

5.4 Understanding relative motion

Lesson overview

Learning objectives

- Describe the motion of objects in relation to each other.
- Explain the concept of relative motion.
- Apply the concept of relative motion to various situations.

Learning outcomes

- Describe a situation in which different objects are travelling at different speeds. [O1]
- Apply the concept of relative motion to a situation with two objects moving at different speeds. [O2]
- Apply the concept of relative motion to a situation with more than two objects moving at different speeds. [O3]

Skills development

- Thinking scientifically: consider the quality of evidence
- Working scientifically: interpret evidence
- Learner development: act responsibly

Resources needed A4 sheets of paper; scissors; paper clips; digital timers; metre rulers or tape measures; graph paper; mini-whiteboards; marker pens; Worksheet 3.5.4; Practical sheet 3.5.4; Technician's notes 3.5.4

Digital resources Quick starter; Interactive activity: Order the relative speeds of the cars, from the fastest to the slowest

Common misconceptions When a free-falling parachutist opens their parachute, they fly upwards.

Key vocabulary relative motion, relative speed

Teaching and learning

Engage

- Explore the students' ideas by showing a video clip of a parachutist when viewed from the ground and when viewed by another descending parachutist. The students **describe** the motion of the first parachutist relative to the ground (accelerating downwards until terminal velocity is reached) and the motion relative to the other parachutist (none). [O1&2]

  Higher-attaining students may be able to **suggest** other situations in which relative motion is important – such as when police are trying to judge the speed of a car they are chasing, or when a driver is overtaking on a single carriageway road.

Challenge and develop

- Ask the students to **consider** the difference between the *absolute* speed of two objects and their *relative* speed. Use the 'Relative motion' section in the Student Book. [O2]
- The students then **make** paper models of sycamore seeds, **test** them and **collect data** of their relative motion, following the instructions on Practical sheet 3.5.4. They will need to drop them from a height of at least 2 m. Tell students to act responsibly and take care to avoid the risk of falling from a height. [O1&2]

  Higher-attaining students may be able to collect additional data to compare the motion of three differently weighted model sycamore seeds. [O3]

- The students design a table to **record** their observations and **plot** a distance–time graph for their model sycamore seeds. They then **interpret** and **consider** the quality of the evidence they have collected and **suggest** what improvements could be made. Practical sheet 3.5.4 can be used to guide the analysis and evaluation of the method and the results. [O2&3]

- Ask them to complete the tasks of Worksheet 3.5.4. They **analyse** the relative motion of vehicles travelling along a single carriageway road. [O2&3]

Explain

- Ask the students to read the 'Journeys and collisions' section in the Student Book. Ask them to **explain** how relative speed is crucially important when considering road traffic collisions. [O1, 2&3]

  Higher-attaining students can be asked to **suggest** steps that could be taken to reduce the risk of damage and reduce the damage caused by head-on collisions. [O2&3]

Consolidate and apply

- Show the students a large traditional analogue clock. Ask them to **compare** the absolute and relative motion of the hands on the clock. Use the linear distance between each of the numbers on the clock face. This requires careful measurement on the clock that is used. [O1&2]
- They **calculate** how long after 12 o'clock it takes for the minute and hour hands to reach the same position again. (Just over 65 minutes because the hour hand moves incrementally just past the '1' as the minute hand reaches the '1' having done one-and-a-bit revolutions.) [O2]

  If there is second hand on the clock, higher-attaining students can consider this in addition to the hour and minute hands. [O3]

Extend

- Ask students able to progress further to **consider** the relative speeds of light (speed ≈ 300 million m/s), sound (speed ≈ 340 m/s), a bullet from a hand gun (speed ≈ 320 m/s) and a bullet from a high-velocity rifle (speed ≈ 900 m/s). If they were standing watching the two guns being fired simultaneously, what would they see, hear and feel, and in what order? [O3]

Plenary suggestions

Mini-whiteboard activity The students draw axes for a distance–time graph on mini-whiteboards and then add lines to show journeys that you describe. Check their responses after each line has been added. [O2&3]

Answers to Student Book questions

1. a) Person will travel 6 km; boat will travel 5 km. b) The jogger is 2 km/h faster than the boat. c) backwards
2. Speed of A = 60 km/h; speed of B = 120 km/h; relative speed of car B to car A is 60 km/h.
3. a) Car A is 2 km ahead of car B, which has not moved from the start point.
   b) Car B has been travelling for 1 min, and has covered 2 km; it is 1 km behind car A.
4. The collision speed is the combined speed of the two vehicles.
5. If you do not have a fixed reference point you do not know if you are moving, or if the object you are looking at is moving, e.g. parallel trains. Sense of balance may be able to help detect movement.
6. Situations a) and b) have the same impact speed, 10 km/h; so the damage from the collisions will be similar.
   b) could be much worse if the driver lost control and hit a stationary object
   c) The impact speed is much higher; 130 km/h; so the consequences would be very serious.

Answers to Worksheet 3.5.4

1. a) car A: 70 m; car B: 80 m b) car A: 7 m/s; car B: 8 m/s
   c) car B is 1 m/s faster. d) 55 m
   e) The additional car is 1 block ahead of car A in picture 1, and 2.5 blocks behind in picture 2.
2. a) Check for correct scales, positions and arrows. If shown travelling left to right, Alex is 200 m from the left-hand end and Georgie is 100 m from the left-hand end.
   b) 500 m from the left-hand end
   c) The cars should be alongside each other 500 m from the left-hand end.
3. a) Asma is at the far right-hand end in the first diagram, and 200 m from the right-hand end in the second diagram.
   b) Not very likely; there is a risk of a head-on collision in just over 3 s.
   c) Overtake as fast as possible; wait for a dual carriageway; be patient; choose not to overtake, because it is unsafe.
   d) Police car is at the far left-hand end in first diagram, and at the far right-hand end in the second.
   e) They need to be able to judge speed of a car and judge risks well. They must not put the public in danger.

5.5 Analysing equilibrium

Lesson overview

Learning objectives

- Analyse situations to identify the various forces that are acting.
- Explore static situations in which objects are held in equilibrium and the nature of the forces involved.

Learning outcomes

- Identify relevant forces and their directions. [O1]
- Explain how opposing forces can result in an object being in equilibrium. [O2]
- Explain how a more complex set of opposing forces can result in an object being in equilibrium. [O3]

Skills development

- Thinking scientifically: use units and nomenclature
- Working scientifically: develop explanations
- Learner development: develop resilience

Resources needed equipment and materials as detailed in the Technician's notes; Worksheet 3.5.5; Practical sheet 3.5.5; Technician's notes 3.5.5

Digital resources Quick starter; Video; Interactive activity: Drag the statements into the correct group - those which describe forces in equilibrium, and those which do not

Common misconceptions No forces are acting if an object is stationary.

Key vocabulary balanced forces, equilibrium

Teaching and learning

Engage

- Show a mass suspended from a newtonmeter. Point out that there is no movement, and ask the students if any forces are acting. They should **identify** the downward force of weight and the upward force through the newtonmeter. [O1&2]
- **Pair talk** The students **explain** what the lack of movement tells them about the forces acting on the newtonmeter. (They are balanced – they are in equilibrium.) [O1&2]

Challenge and develop

- The students **analyse** a simple equilibrium situation, by reading the 'Forces in balance' section in the Student Book and answering questions 1–3. [O1&2]
- Ask the students to **draw** the force arrows acting on a piece of soft foam that is stationary on a bench and to **explain** why there is no movement. (The upward force from the bench is in equilibrium with the downward force of the foam's weight.) They **discuss** how the force diagram would be different if a large mass were placed on top of the foam. (The bench would provide a bigger upward force, which would be in equilibrium with the combined forces of the foam's weight and that of the added mass.) [O1&2]
- **Pair work** The students work in pairs to exert opposite pulling forces on a heavy object on the floor, using newtonmeters. See Technician's notes 3.5.5 and Practical sheet 3.5.5. The students **consider** why the two newtonmeters can pull with different forces and yet the object does not move. (The force of friction added to the smaller force balances the larger force.) [O2&3]

  Higher-attaining students may be able **explain** why equilibrium can exist for a range of different pulling forces. (Friction opposes movement – so long as the force of friction is not overwhelmed it will resist any movement.)

Explain

- The students **consider** a variety of equilibrium situations, e.g. a bulldog clip holding paper, a gymnast balancing on a beam, washing hanging on a line, and **explain** how the forces balance each other. [O1, 2&3]

  Higher-attaining students may **suggest** other situations in which more than two forces are acting and are in equilibrium.

Consolidate and apply

- Ask the students to **compare** the forces acting when a heavy object is on a rigid beam to when the object is on a flexible beam. (They should identify that the forces are the same in both cases.) [O1&2]
- **Pairs to fours** Ask the students to **compare** the forces acting when an object is floating in water to when some extra weight is added so that the object sits slightly lower in the water. (The weight and upthrust forces balance in both cases; the forces are higher in the second situation.) [O2&3]
- The students **analyse** equilibrium through the tasks of Worksheet 3.5.5. [O1, 2&3]

Extend

- Ask students able to progress further to **draw** diagrams, with force arrows to scale, to show the forces acting in more complex equilibriums, for example those in task 3 of Worksheet 3.5.5. They will need to **show care** and **resilience** to succeed with the scale diagrams. [O3]

Plenary suggestions

The big ideas Ask the students to write down three ideas they have learned during the lesson. Then ask them to **share** their facts in a group and to **compile** a master list of facts, with the most important at the top. Ask for ideas to be shared to find out whether groups agreed or not. [O1, 2&3]

Answers to Student Book questions

1. downward force of weight; upward force provided by the spring
2. first diagram: weight arrow acting downwards; equal and opposite force provided by the spring
   second diagram: as for the first diagram but both arrows should be larger
3. The forces balance; they are in equilibrium.
4. The large force arrow shows pushing from the left; two smaller force arrows – pushing force and friction – are acting in the opposite direction. Higher-attaining students should draw the arrows to scale to show that they balance.
5. The effect of the force of friction is reduced, so the forces remain in balance.
6. a) Air resistance and friction are acting to the left. b) 1150 N c) 1150 N d) The forces are in equilibrium.
7. Any suggestions to reduce air resistance, e.g. air deflector on car roof; or to reduce friction, e.g. lubricating moving parts.

Answers to Worksheet 3.5.5

1. a) All the forces acting on an object balance one another.
   b) i) The downward force of the person's weight; the equal and opposite upward force from the mattress springs.
   ii) The downward force of the person's weight; the equal and opposite upward force from the scales.
   iii) The downward force of the coat and hanger's weight; the equal and opposite upward force from what the hanger is attached to.
2. a) The downward force of the weight of the person; the equal and opposite upward force from the beam; the downward force of the weight of the beam; the upward forces from the supports at either end of the beam; balancing the weight of the beam and the person.
   b) There would be a larger downward force of the weight of the person; also larger downward forces exerted on the supports of the beam; but larger upward forces to balance the extra weight.
3. a) The forces of air resistance and friction are in equilibrium with the pushing force from the engine.
   b) Friction from the tyres and/or the force through the steering wheel, to balance the pushing force of the wind.
   c) There is a large sideways force caused by air resistance against the side of the vehicle. It could cause the vehicle to swerve off course or to be blown over.

Answers to Practical sheet 3.5.5

1. friction between the box and the floor 2.The arrow for the friction force is in the opposite direction to the pulling force and is the same size. 3.The first pulling force is now greater than the friction force, but there is a pulling force backwards, of length equal to the difference between the forward pulling force and the friction. 4.The forces are balanced: the total forward force is equal to the total backward force.

5.6 Exploring motion and equilibrium

Lesson overview

Learning objectives

- Explore dynamic situations which may involve equilibrium.
- Apply ideas about equilibrium to a situation in which an object is moving.

Learning outcomes

- Explain how forces can cancel each other out. [O1]
- Explain various situations in which objects may or may not be in equilibrium. [O2]
- Explain more complex situations in which objects may or may not be in equilibrium. [O3]

Skills development

- Thinking scientifically: ask questions
- Working scientifically: make predictions
- Learner development: communicate effectively

Resources needed paper spinner from Topic 5.4; equipment and materials as detailed in the Technician's notes; Worksheet 3.5.6; Practical sheet 3.5.6; Technician's notes 3.5.6

Digital resources Quick starter; Slideshow: Forces in sky-diving - From free-fall to an open parachute; Interactive activity: Match the sentences about motion and equilibrium; Hangman: Key vocabulary game

Common misconceptions Objects stop moving when the force runs out.

Key vocabulary equilibrium, unbalanced forces

Teaching and learning

Engage

- Using the paper spinner from Topic 5.4, demonstrate the motion of a sycamore seed falling from a tree. The students **describe** the motion and forces carefully using mini-whiteboards. (The spinner accelerates downwards and starts spinning. It then descends at a constant speed. The forces acting are weight, gravity and air resistance.) Ask the students to read the 'Forces in equilibrium' section in the Student Book and to answer questions 1–3. [O1&2]
- Then ask the students to **draw** a force diagram for a sycamore seed descending at a steady speed. (The downward force arrow due to weight is balanced by an equal and opposite force of air resistance.) [O1&2]

Challenge and develop

- Demonstrate movement at a steady speed using a glider on a linear air track. Encourage the students to **ask questions** about the similarities and differences between this situation and the descending sycamore seed. (The descending seed has two forces in equilibrium. Once the glider is moving it has no forces causing acceleration or deceleration. In both cases there is steady speed.) [O1&2]

  Higher-attaining students may also **consider** the downward force of gravity and the weight of the glider, and the equal and opposite force of the air that supports it.

- Demonstrate the glider on the air track – this time with an elastic band at each end. After moving backward and forward a few times, the glider will slow and stop. Working in pairs, the students **discuss** what this shows. (The forces on the glider are not quite in equilibrium because there is a small amount of friction present.) [O1&2]
- The students **discuss** how two forces acting in the same direction can be added together – for example someone pedalling a bicycle and someone pushing from behind – or the forces can be subtracted when they are acting in opposite directions – for example someone pedalling a bicycle and someone pushing against them. [O2&3]

- The students **investigate** the motion and equilibrium of a dynamics trolley. Practical sheet 3.5.6 gives details of the experiment. They need to set up a friction-compensated ramp (a slope just steep enough for the trolley to roll down at a constant speed). [O2&3]
- The students read the second and third sections of the Student Book and answer the questions. [O2&3]

Explain

- **Group talk** The students work together to **explain** how an ice-hockey puck can travel from one end of the rink to the other without appearing to slow down. They should account for gravity in their explanation and a small amount of friction. Selected groups **communicate** effectively to **present** their explanations to the class. Ask other students to feedback three positive points and one 'even better if' point. [O1, 2&3]

  Higher-attaining students may be able to **relate** the behaviour of an ice-hockey puck to that of the glider on the air track.

Consolidate and apply

- Show the students a dynamics trolley with a marble fixed on top of it in a Plasticine 'seat'. Ask them to **predict** what will happen to the marble if the moving trolley hits an obstacle and comes to a sudden halt. (The marble should continue forward when the trolley stops – demonstrate this for the class.) The students **write explanations** of the outcome, referring to the forces acting on the trolley but not the marble. [O2&3]
- Challenge any misconceptions using tasks 1 and 2 of Worksheet 3.5.6. [O2&3]

Extend

- Ask students able to progress further to **predict** and **explain** what would happen to a marble if it were rolled down two different ramps (task 3 of Worksheet 3.5.6). [O3]

Plenary suggestions

I think that... Working in pairs, students **summarise** the key learning points. Some may benefit from sentence stems such as 'An object moving in a straight line at a steady speed may have no forces acting on it but...'. [O1, 2&3]

Answers to Student Book questions

1. stays the same 2. Gravity is acting downwards; air resistance is equal and opposite. They may comment that air resistance is lower initially. 3. Because the forces acting on it become balanced; and are in equilibrium.
4. Forces that do not cause a change in the movement of an object, because they cancel out each other's effects.
5. The pushing force of the train engine is balanced by the equal and opposite friction force, including air resistance. The downward force of weight is balanced by the equal and opposite upward force from the rails.
6. If the wind increases, the air resistance is greater than the pushing force from the engine; so the driver must push on the accelerator; to increase the pushing force of the engine; the opposite happens if the wind drops. They may comment on gusts from the side needing to be balanced by a steering force; or the fact that perfect equilibrium is hard to maintain.
7. A car experiences a force when it hits an object, so it slows and/or changes direction. The passengers, unless wearing seatbelts, will continue at the same speed and direction, until they hit the dashboard, windscreen or another object. Scientific ideas that could be involved include: balanced forces before the collision; unbalanced during the collision; seat belt absorbing some of the shock. Other possible examples include: an object being fired from a catapult; an egg breaking when it hits the floor.

Answers to Worksheet 3.5.6

1. a) Gravity (weight) pulling down; air resistance is equal and opposite; they are in equilibrium.

   b) The ball pushing the net; the net pushing back on the ball; gravity pulling down; they are not in equilibrium.

   c) The pushing force through the pedals; equal and opposite air resistance and friction; gravity and weight pulling down; equal and opposite force provided by the road; they are in equilibrium.
2. Statement (c) is correct; they could add further explanation, such as what effect air resistance has.

   The first part of (b) is correct but the second part is wrong; as the pendulum moves upwards, gravity and weight act downwards, so it slows down rather than continues up at the same speed.

   Statement (a) is wrong, because the pendulum is changing direction, so a force must be acting (gravity). Even if the pendulum was stationary, forces can still be acting but be in equilibrium.
3. a) In the first case the ball goes to same height on the other side; gravity/weight causes it to accelerate downwards, and then slow to a stop on the other side. In the second case the ball keeps going along the horizontal surface forever.

   b) In the first case the force of friction opposes movement, so the ball will not go to the same height. In the second case the ball will slow down and eventually stop.

   c) The pushing force would be added to the downward force of weight/gravity. In the first case the marble would fly upwards off the ramp. In the second case the marble would go faster than in a), and faster/further than in b).

5.7 Applying key ideas

Crossing the English Channel

Objectives

- Analyse distance–time data.
- Interpret data and apply ideas about motion to the context of ferry and train travel.
- Use ideas about forces, including equilibrium, to compare and explain various aspects of motion.

Outcome

- Making clear and effective responses to questions, indicating understanding and the next steps in learning.

The purpose of this activity is to provide an opportunity to see how successfully the students are grasping the key ideas so far. It is not designed to be used as a formal test – it might be that the students work on the tasks collectively. It provides an opportunity for you to look at written work, engage students in discussions and form ideas about progress being made.

The tasks are progressive. Lower-attaining students should be able to tackle the first two or three tasks and higher-attaining students will find the later tasks more challenging.

Resources needed graph paper; rulers; pencils

Teaching and learning

Engage

- Challenge the students to **reflect back** over the last few lessons and **identify** some of the key ideas they have met. Draw out some responses about these ideas and say that it is important as learners that they develop a sense of their own progress.
- Tell the students about the two different options for taking a car across the English Channel – ferry or tunnel, where the cars drive onto a train and are carried to the other side.

Challenge and develop

- **Pair work** The students work in pairs to **sketch** a distance–time graph to show what crossing the English Channel by ferry may look like. (The graph should show a period of acceleration where the line curves upwards, a straight upward slope for steady speed, and then level out at the end of the journey.) Ask the students to **describe** how the graph would change in different scenarios, such the ferry giving way to another ship or a strong tail wind during the crossing.
- Ask the students to **brainstorm** all the different forces that could be acting during the crossing, such as the pushing force from the engines, air and water resistance, weight/gravity and upthrust.

Explain

- Ask the students to read 'Crossing the English Channel' in the Student Book. Then ask them to **identify** contexts met in previous lessons that are similar to the one here. Using the text as a source of ideas, draw a spider diagram on the board with 'motion' at the centre and the students' ideas around it.

Consolidate and apply

- Now ask the students to attempt the tasks in the Student Book – they might do this individually or collaboratively. In either case, encourage them to **identify** and **record** their ideas. Ask them to **present** their responses to various tasks, either orally or by displaying their work. Depending on their level of independence, you may need to adopt strategies such as:
  - providing some students with axes/ scales for plotting a graph
  - reminding students how to draw force arrows and calculate speed.

- working with groups of students who have difficulty with longer answers, to help them **plan** and **structure** their responses.

Extend

- For some students task 6 gives an opportunity for extension work.

Plenary

The big ideas Ask the students for responses to questions such as:

- How can you make sure that you include enough detail in your responses?
- Are there some ideas that you need to strengthen in the later part of this chapter?

Provide the class with some overall feedback, indicating ideas that they have grasped effectively and those that may need developing further.

Likely responses and next steps in learning

1. This task gives students an opportunity to reflect on their graph-plotting skills. Most should be able to: decide on suitable scales – linear, filling at least half the space, with sensible intervals such as fives or tens; fully label axes including units; accurately plot points in sharp pencil to the smallest division on the graph paper or smaller; draw appropriate straight lines and smooth curves; add a title; label each curve/ line.
2. This task encourages students to consolidate and embed their graph-plotting techniques. They should demonstrate this by including all the points listed in task 1. They could also be given the opportunity to peer-assess their graphs.
3. In this task, answers will depend on the accuracy of their graph, so they should not be penalised twice for the same error. After 25 minutes the ferry will be about 6.5 km into the journey, although they are reading from a curved part of the graph which may introduce some additional error. The train will be 34 km into the journey.

   The mean speeds are: ferry = 28 km/h; train = 78 km/h. The gradients indicate speed: the steeper the line, the faster the train or ferry is travelling. A horizontal section indicates that a vehicle is not moving.
4. The mean speed is 4.5 km/h. The line added to the graph should start at point (0, 0) and pass through point (120, 9). Speed may vary because: tides and currents may affect the speed of the swimmer; they may stop to rest; at times they will swim faster, at other times slower.
5. The students should draw a series of diagrams that include force arrows in proportion across the series. They should clearly show how horizontal forces are not in equilibrium when the ferry is speeding up or slowing down. Generally the vertical forces of upthrust and weight/gravity are in equilibrium, but higher-attaining students may be encouraged to think about the influence of waves and the gradual loss of weight of the ship as it burns fuel.
6. In task 5, and everywhere else, Newton's First Law applies. If a car is not fastened when the train crashes and stops or slows down suddenly, the car will carry on and crash into the first stationary object is meets. However, the force of friction between the car tyres and the surface underneath means that the car will not carry on at the same speed as before the crash; it will slow down. In the case of a gentle crash, the force of friction may be sufficient to stop the car moving. People would react in a similar way to the car – if they are wearing seat belts, they will experience whatever force/ motion the car does; if they are walking around they will tend to carry on forward when the train stops until they hit something. There is also the added danger of them coming to rest and then being hit by other objects that are still moving.

5.8 Understanding gravitational fields

Lesson overview

Learning objectives

- Describe gravity as a non-contact force.
- Explore the concept of gravitational field and weight.
- Relate this concept to life on Earth.

Learning outcomes

- Describe the effect of an object being in a gravitational field. [O1]
- Explain the relationship between gravity and weight. [O2]
- Explain weight in relation to the idea of a gravitational field. [O3]

Skills development

- Thinking scientifically: use equations
- Working scientifically: develop explanations
- Learner development: respect others

Resources needed flip chart paper; marker pens; calculators; top-pan balance; beaker of water; newtonmeter; rubber bung; Worksheet 3.5.8

Digital resources Quick starter; Video; Interactive activity: Order the weights, from largest to smallest; Slideshow: Mass and weight - An explanation of the two

Common misconceptions Heavy objects fall faster than lighter objects. Gravity is linked to the Earth's rotation.

Key vocabulary gravitational field, gravitational field strength, weight, mass

Teaching and learning

Engage

- Identify the students' prior knowledge of different types of fields by asking them to work in groups of four to **record their ideas** on flip chart paper, which they **share** with others. Some may need prompts – different types of fields, characteristics, similarities and differences, how they can be detected or measured. [O1&2]
- Display the students' main ideas about gravitational fields, including any vague or incorrect ones, with a view to respecting all suggestions and to accepting, modifying or rejecting them later in the lesson. [O1&2]

  Higher-attaining students may comment on the relationship between gravity and weight.

Challenge and develop

- **Pairs to fours** The students **describe** whether or not the strength of the gravitational field of a planet is uniform, and if not how it varies. (It depends on the mass of the planet and distance from the planet.) Emphasise that the strength of a gravitational field of a planet does not depend on the presence or absence of other objects within the field. Task 1 of Worksheet 3.5.8 can support this activity. [O1&2]
- Discuss why different objects have different weights, even though the gravitational field they experience is the same. (Weight depends on mass.) Check that the students understand what 'mass' is (a quantity of matter: number and type of particles) and that gravitational fields are a pulling force. [O1, 2&3]
- Introduce the formula that links weight and mass (see the Student Book) – weight = mass × gravitational field strength ($W = mg$). ('Weight' and 'gravitational force' are the same, which is why weight is measured in newtons (N).) [O3]
- Students **calculate** gravitational field strength using an example object with a mass of 10 kg that weighs 100 N, and by substituting the values and solving the equation. (100 N = 10 × g, so g = 10). From a

rearranged equation, they **identify** the correct unit for gravitational field strength. (g = weight ÷ mass, so g is measured in N/kg) [O2&3]

Explain

- Using the formula for weight ($W = mg$) and the value for gravitational field strength (g) of 10 N/kg, the students **explain** why objects of different masses have different weights on Earth (task 2 of Worksheet 3.5.8). [O1, 2&3]

  Higher-attaining students may be able to **explain** how differences in the weight of an object on different planets can be linked to the gravitational field strength of each planet.

Consolidate and apply

- The students use the idea of gravitational field to **explain** why, when measuring weight, it is necessary to specify where the measurement is taken – for example on the *surface* of *the Earth*. [O2&3]
- **Pairs to fours** The students **compare** the weights of a rubber bung in air and in water (demonstration will suffice) and **suggest** how this might relate to the ideas of weight and mass. (Weight is a force that can vary; mass is fixed.) [O2&3]

Extend

For students able to progress further:

- They **consider** that 'acceleration due to gravity' is due to the gravitational field strength and they have the same value. So at a location where the acceleration due to gravity is 10 m/s^2, the field strength is 10 N/kg. They can now do task 3 of Worksheet 3.5.8. [O3]
- The students **suggest** what equipment and method would be needed to measure the gravitational field strength at a particular location. (They may suggest time-lapse photography or light gates to measure speeds and calculate acceleration.) [O3]

Plenary suggestions

Learning triangle Ask the students to draw a large triangle with a smaller inverted triangle that just fits inside it (so they have four triangles). In the outer three, ask them to think back over the lesson and write in the respective triangles something they've seen, something they've done and something they've discussed. Then they add something they've learned in the central triangle. [O1, 2&3]

Review Go back to the students' statements from the *Engage* activity and discuss how they might be amended in the light of new learning. [O1, 2&3]

Answers to Student Book questions

1. The space around an object in which it pulls other objects towards it.
2. Sea levels on Earth rise and fall with the tides.
3. Gravity is a non-contact force; a tug-of-war involves a contact force.
4. The mass of the object and the gravitational field strength at its location.
5. The amount of matter present; the number and type of particles in the object.
6. Gravitational field strength, which varies with distance and location, determines how strongly an object is pulled; its weight. The number and type of particles (mass) in the object are fixed.
7. Weight is defined by $W = mg$, where g depends on gravitational field strength; a stronger gravitational field leads to a bigger weight.
8. Acceleration due to gravity and gravitational field strength are equivalent, so have the same value. The field strength does not depend on the object; it depends on the object's position in relation to the Earth.
9. Drop two spherical objects of equal mass but different densities, so one has a larger size. If they hit the ground at different times, air resistance is an influence, because all other variables are controlled. They may mention measuring techniques and repeating trials.

Answers to Worksheet 3.5.8

1. a) False; weight depends on the strength of the gravitational field. b) True c) False; mass depends on the number and type of particles in an object. d) True e) False; it extends well into space, beyond the atmosphere.
2. They should make correct reference to matter/particles, gravitational field strength, and the formula $W = mg$.
3. They should refer to the formula $W = mg$ to show how weight depends on mass. The mouse weighs 1 N, which is the force it is pulled down by. The elephant weighs 50 000 N; both accelerate due to gravity at 10 m/s^2.

5.9 Applying ideas about gravitational fields

Lesson overview

Learning objectives

- Apply the concept of gravity causing weight to other situations.
- Explore the implications of varying gravitational field strength.

Learning outcomes

- Describe how weight varies near to different bodies. [O1]
- Explain why weight changes near to other bodies. [O2]
- Use the concept of gravitational fields in relation to various bodies. [O3]

Skills development

- Thinking scientifically: understand how theories develop
- Working scientifically: make predictions
- Learner development: plan progress

Resources needed range of spherical objects of different diameters and masses to represent planets e.g. football, balloon, French boule, tennis ball, marble, polystyrene ball; Worksheet 3.5.9

Digital resources Quick starter; Interactive activity: Drag the statements into the correct group - those which describe a strong gravitational field, and those which describe a weaker field

Common misconceptions Objects are always weightless in space.

Key vocabulary gravitational field

Teaching and learning

Engage

- Ask the students to **speculate** on the path that a rocket would take if it was fired from Earth into space towards two distant planets. (Task 2 of Worksheet 3.5.9 provides a framework for this, and can be done now on the board or can be completed later in the lesson by the students.) The students **discuss** the possibilities and **identify** additional information that would help them to **make predictions** – for example the strength of the gravitational fields of the different planets that the rocket would experience. [O3]
- Ask the students to **predict** and **explain** the reading that the astronauts would see if they weighed themselves on bathroom scales once the rocket's engines had been turned off – this is question c) in task 2 of Worksheet 3.5.9. (The astronauts would be weightless because they and the rocket are subject to the same gravitational field.) [O3]

Challenge and develop

- Referring to the 'Field strengths of planets and stars' section in the Student Book, discuss why the radius of a planet affects its gravitational field strength and the weights of objects on the surface of the planet. [O2]
- The students **predict** how someone's weight would differ at sea level compared to at the top of Mount Everest. (The person would be lighter on Mount Everest because they would be further from the Earth's centre.) [O1, 2&3]

  Higher-attaining students can be introduced to the quantitative relationship: $g \propto 1/r^2$ where *g* is the gravitational field strength and *r* is the distance from centre of the planet.

- The students **predict** how someone's weight would vary over the surface of a perfectly spherical planet compared with over the surface of a planet that is flattened north to south. (They would have the same weight everywhere on a spherical planet, and a heavier weight near the poles of the flattened planet.) [O1, 2&3]

  Higher-attaining students may **consider** how a varying density of the inside of a spherical planet will affect gravity.

Explain

- The students look at and hold a number of spherical objects that represent planets (see *Resources needed*). They **predict** a rank order for a person's weight on the surface of each planet. (Weight reduces with planet radius and increases with planet mass.) They **identify** the next steps, in terms of additional information they would need, to **improve** their predictions and **explain** why. (The quantitative relationship of how gravitational field strength varies.) [O1, 2&3]

Consolidate and apply

- **Pairs to fours** The students should test each other by **devising** their own problems about weight and gravity. [O1, 2&3]
- They attempt the tasks of Worksheet 3.5.9 to strengthen their knowledge of the topic and to **think more broadly** about the work of scientists. [O1, 2&3]

Extend

- Ask students able to progress further to **consider** Mercury's orbit, which (like Uranus) did not follow the expected orbit. Originally it was proposed that the gravity from another, yet-to-be-discovered, planet was affecting Uranus. Scientists became convinced that the proposed planet did not exist and came to accept that things can happen that do not fully obey Newton's Laws. The students **suggest** how this could have been a stimulus for the development and testing of Einstein's Theory of General Relativity. [O3]

Plenary suggestions

True or false? The students work in small groups to **devise** a 'true/false' quiz to test each other's knowledge and understanding of the lesson. [O1, 2&3]

Answers to Student Book questions

1. You are further from the planet, so the gravitational field strength is smaller.
2. The one closer to the surface of the Earth would speed up more quickly (greater acceleration), because it experiences a higher gravitational field strength.
3. The field strength would be stronger from the Earth; it has a bigger mass than the Moon.
4. The small planet would have a higher density.
5. Lines radiate out from the centre point of the planet. Where the lines are close together the field strength is bigger, so weight is also greater.
6. Newton's theories suggested that gravity from an unseen object was affecting the orbit of Uranus.
7. Theories need to be continually tested to see if they can correctly predict events/phenomena. If the observations prove the predictions wrong, the theory needs to be modified or rejected.
8. The new planet has a larger radius than Earth.

Answers to Worksheet 3.5.9

1. a) i) no effect

   b) D

   c) Greater than; because its radius is smaller.

   d) A large radius tends to lead to a weaker surface gravity; so for D to have the highest surface gravity, it must have a high mass/density.
2. a) The gravitational field of the Earth gets weaker, the further it travels from the Earth. There is little or no influence at long distances; the rocket may experience the gravitational fields of other planets, which get stronger, as it gets closer.

   b) The masses of the planets; the gravitational field strength of the planets.

   c) During launch their weight would be greater. Once the rocket engines were turned off, the astronauts' weights would be zero, because any pull of gravity would affect them, the rocket and all its contents equally.
3. Reward any reasonable suggestions along the lines of the examples below.

   a) The observations provide evidence to support the theory.

   b) They can check the evidence and conclusions; they offer a fresh outlook; they can challenge the theory.

   c) If it is providing correct explanations for phenomena, it should be able to correctly predict related phenomena. If it makes incorrect predictions, the theory may be fully or partially incorrect.

5.10 Looking at motion in the Solar System

Lesson overview

Learning objectives

- Relate ideas about gravitational fields to the Sun–Earth–Moon system.
- Use these ideas to explain the position and motion of these bodies.

Learning outcomes

- Describe how the Sun, Earth and Moon move in relation to each other. [O1]
- Use the concept of gravity to explain how the Sun, Earth and Moon move in relation to each other. [O2]
- Analyse the motion of objects in gravitational fields. [O3]

Skills development

- Thinking scientifically: understand how theories develop
- Working scientifically: make predictions
- Learner development: respect others

Resources needed poster paper; coloured pens; rubber bung on a string; Worksheet 3.5.10

Digital resources Quick starter; Interactive activity: Order the planets of the Solar System, from the furthest from the Sun to the nearest; Slideshow: Orbital motion - A look at how we put satellites into space

Common misconceptions If the force holding an object in circular motion is removed, the object will fly away at 180° to the force.

Key vocabulary orbit, ellipse, asteroid

Teaching and learning

Engage

- If possible, show the first 2 min 45 s of the video 'Models of the Solar System – Earth, Sun and Moon' from the Institute of Physics website – go to www.iop.org and search for 'Teaching astronomy and space videos', then click on the first link in the list of results. The students **identify** how current theories developed from those of the Ancient Greeks. They should **show respect** for early theories and why understanding the movement of the planets and stars has proved difficult (inaccurate observations, incorrect models, everything is moving). [O1]

  Higher-attaining students may be able to appreciate that the rotation of the Earth and the rotation of the stars could appear the same to an observer on Earth.

Challenge and develop

- Ask the students to **list observations** or **evidence** which suggest that gravitational fields play a part in the movement of the planets. (The orbits follow a pattern; if no forces act then everything travels in straight lines or remains still.) [O2]
- The students **explain** what holds an object in its circular path when it is swung round on a string. Demonstrate this with a rubber bung on a string, in an open space and ensuring that students are at a safe distance. They **link** the pulling force of the string to the pulling force in a gravitational field. [O2&3]
- **Pair talk** The students use ideas from Topic 5.6 to **predict** what would happen to the circular motion if the string snapped or the gravitational field suddenly disappeared. (The object would continue in a straight line at a tangent.) [O2&3]
- The students watch the remainder of the video (2 min 45 sec to 7 min 16 sec) and **suggest** why the planets in the Solar System orbit the Sun rather than the Earth. (The Sun has the largest mass and the strongest gravitational field.) [O2&3]

Explain

- **Group work** The students **create** an annotated poster to show the motion of the planets in the Solar System. They include information boxes to outline the role of the gravitational field. [O1, 2&3]

Consolidate and apply

- The students complete tasks 1 and 2 of Worksheet 3.5.10. [O1, 2&3]
- Ask them to read about the motion of moons in the Solar System in 'The effect of gravitational fields' section in the Student Book, and to answer questions 4–6. Discuss possible reasons why different planets have varying numbers of moons (question 5). [O2&3]
- Ask the students to **describe** how artificial satellites are kept in orbit. Compare their ideas with the 'Forces, speed and orbits' section of the Student Book. [O2&3]

Extend

- Ask students able to progress further to complete task 3 of Worksheet 3.5.10. This includes **researching** and **explaining** the motion of Foucault's pendulum (in the London Science Museum). [O3]

Plenary suggestions

What would happen if…? The students suggest what would happen in the Solar System if:

- a star the size of the Sun passed close to the Solar System. (The planets' orbits would be disrupted or destroyed.)
- the Sun's gravitational field did not exist. (Planets would travel in straight lines or have curved paths influenced by other gravitational fields.)
- the Earth's gravitational field did not exist. (The Moon would drift off until it was attracted by the gravity from other bodies. The orbits of other planets might change slightly.) [O1, 2&3]

Answers to Student Book questions

1. Should show an ellipse for Mars, with the Sun in the centre. The Earth's orbit is inside that of Mars, and is less elliptical.
2. The planet would be colder when it is further away from the Sun.
3. Mercury; Venus; Earth; Mars; Jupiter; Saturn; Uranus; Neptune
4. Planets orbit the Sun; moons orbit a planet as it orbits the Sun.
5. More massive planets tend to have more moons, although this is not always the case. More mass means there is a larger gravitational field to hold objects in orbit. Other factors include the gravitational fields of other bodies, and chance events such as the presence or absence of material to form moons.
6. Its large mass means that it has a far-reaching gravitational field.
7. Its mass, speed and height above the planet, and the strength of the planet's gravitational field.
8. The one above Saturn would need to be further out from the centre of the planet, or be travelling faster, because Saturn's larger mass results in a stronger gravitational field. If the satellite was not high/fast enough, it would fall to the planet.
9. Jupiter's large mass means that it has a strong gravitational field that attracts debris over great distances. This is due to its gravity; whereas a vacuum cleaner uses a reduction in pressure to attract dust.

Answers to Worksheet 3.5.10

1. a) Mercury; Venus; Mars; Saturn

   b) The longer the distance from the Sun, the longer the orbit time and the lower the mean orbit speed.

   c) Saturn has the strongest gravitational field (due to its larger mass); its gravitational field extends further.
2. a) Check that accurate diameters are drawn. b) It has a similar size to some larger asteroids.

   c) 108 cm d) Some curvature should be visible.
3. a) Its plane of swing changes in a regular and predictable way as the Earth spins beneath it.

   b) At the poles it is on the axis of rotation, where that point of the Earth spins on the spot; elsewhere it is orbiting some distance from the axis of rotation.

   c) He proposed that the Sun is at the centre of the Solar System.

   d) He used a telescope for astronomical observations, so was able to make many observations that had not been possible before.

5.11 Describing stars and galaxies

Lesson overview

Learning objectives

- Describe the characteristics of a star.
- Relate our Sun to other stars.
- Explain the concept of galaxies and the position of our galaxy compared to others.

Learning outcomes

- Describe the differences between the Sun, other stars and galaxies. [O1]
- Describe the relationship between the Sun, other stars and galaxies. [O2]
- Relate ideas about the Sun, stars and galaxies to evidence visible from the Earth. [O3]

Skills development

- Thinking scientifically: ask questions
- Working scientifically: develop explanations
- Learner development: ask questions

Resources needed laptops or a computer suite; Worksheet 3.5.11

Digital resources Quick starter; Interactive activity: Match the statements about stars, galaxies and the Universe

Common misconceptions Our Solar System is at the centre of the Universe.

Key vocabulary star, nuclear fusion, red giant, neutron star, galaxy

Teaching and learning

Engage

- Identify the students' prior knowledge by asking them to work in small groups to **compile a fact file** about the Sun. Possible prompts to provide them with are shape, size, energy source, origin, composition, differences to planets. [O1]
- Discuss the students' fact files and address any misunderstandings. Provide additional information such as diameter (1.3 million km) and composition (75% hydrogen plus helium and other elements). [O1]
- The students **ask questions** about things that they would like to find out about the Sun, stars and galaxies. (They will research the answers to these later in the lesson.) [O1, 2&3]

Challenge and develop

- The students read the first two sections of the Student Book and **deduce** why stars have a finite lifespan. [O1&3]
- If possible, show the video 'The Life Cycle of Stars' (7 minutes) from the Institute of Physics website – go to www.iop.org and search for 'Teaching astronomy and space videos', then click on the first link in the list of results. The students **make comparisons** between the different types of stars in terms of size, energy source and appearance. [O1&3]
- They **produce a summary report** on star or galaxy observations from the National Schools' Observatory website. To do this go to www.schoolsobservatory.org.uk and click on 'Register your School'. [O1, 2&3]

  Higher-attaining students may be able to **undertake a longer-term project** using the National Schools' Observatory.
- Students **compare and contrast** stars they have researched. [O1, 2&3]

- They **research** some of the methods of gathering information about stars and galaxies – telescopes, measuring different types of light/radiation, measuring rates of expansion of galaxies. [O3]

Explain

- The students **describe** the similarities and differences between the Sun and other stars and then **explain** the reasons for these. [O1&2]
- They **explain** how the Solar System would be different if the Sun was a much bigger star. [O1&2]

Higher-attaining students may be able to **speculate** on the likelihood of us detecting life elsewhere in the Universe.

Consolidate and apply

- The students **explain** why the age, type and distance of the nearest star are critical factors for life to exist on a planet. [O1&2]
- They **research** answers to the questions they identified earlier in the lesson. [O1, 2&3]
- They then complete the tasks of Worksheet 3.5.11. [O1, 2&3]

Extend

- Ask students able to progress further to, if possible, watch the video 'The Sun' from the Institute of Physics website – go to www.iop.org and search for 'Teaching astronomy and space videos', then click on the first link in the list of results. Then ask the students to **explain** how fluctuations in the Sun's activity are monitored and why they are important to us. [O3]

Plenary suggestions

Glossary The students **compile** a glossary of key words relating to the Sun and other stars. [O1, 2&3]

Answers to Student Book questions

1. hydrogen (H) and helium (He)
2. nuclear fusion of hydrogen atoms to make helium
3. The *Sun* (a star) is continually releasing massive quantities of energy, including heat and light from nuclear fusion reactions.

   Planets release negligible amounts of energy by comparison, e.g. from volcanoes, but have no naturally occurring nuclear fusion reactions.
4. Very strong, because of its huge mass, and the radius is small.
5. The energy source of hydrogen for nuclear fusion runs out. It can carry out fusion with helium and some other elements, but they too eventually run out.
6. For convenience, when the numbers are extremely large (or small).
7. Planets are the smallest and orbit stars. Several billion stars exist in a group called a galaxy. Many galaxies make up the Universe. The Universe includes all matter.
8. Ideas such as: peer review; making and testing predictions; collecting further data; and comparing it to the original data.

Answers to Worksheet 3.5.11

1. a) nuclear fusion; atoms being forced together to form new atoms and release energy

   b) star; huge mass of hydrogen and other elements, releasing huge quantities of energy by nuclear fusion

   c) red giant; star that has run out of hydrogen and is fusing helium atoms to release energy, and has expanded
2. a) visible images from telescopes; X-rays; cosmic microwaves

   b) The moment when the Universe formed in a massive explosion from a highly concentrated starting state.

   c) It has been continually expanding.

   d) Models help us to picture or understand objects or events that are very large, small or complex. Models can be compared and tested against reality; and then accepted, improved or rejected.
3. a) Hubble's view is not spoiled by light, dust, water vapour, other matter, or interference in the Earth's atmosphere.

   b) Hubble can see further than was possible before, allowing us to see very dim light from distant ancient galaxies. Spots of light that were thought to be single stars or clouds of gas have now been shown to be whole galaxies of stars. Light from the most distant galaxies has been travelling since shortly after the Universe formed in the Big Bang; it has given a view almost as far back as then.

   c) Hubble's mirror was fractionally the wrong shape, leading to images that were blurred. Astronauts have now fitted equipment to correct the fault, just as spectacles correct faulty vision.

5.12 Explaining the effects of the Earth's motion

Lesson overview

Learning objectives

- Describe variation in length of day, apparent position of the Sun and seasonal variations.
- Compare these with changes in the opposite hemisphere.
- Explain these changes with reference to the motion of the Earth.

Learning outcomes

- Describe the implications of the Earth being tilted on its axis. [O1]
- Explain changes to days and seasons in different hemispheres. [O2]
- Explain what would happen if the Earth's axis was tilted by a different amount. [O3]

Skills development

- Thinking scientifically: use units and nomenclature
- Working scientifically: develop explanations
- Learner development: ask questions

Resources needed equipment as detailed in the Technician's notes; Worksheet 3.5.12; Practical sheet 3.5.12; Technician's notes 3.5.12

Digital resources Quick starter; Video; Slideshow: The effects of rotation and tilt - The Sun shines by day, and the moon shines by night ... don't they?; Interactive activity: Reorder the sentences to explain the effect of the Earth's tilted axis

Common misconceptions Seasonal changes are due to the varying distance of the Earth from the Sun.

Key vocabulary axis of rotation, season

Teaching and learning

Engage

- The students answer a starter quiz to establish prior knowledge (task 1 of Worksheet 3.5.12). Make a mental note of their responses that indicate gaps in their knowledge. [O1&2]

  You could ask some students to **review** their own 'starting points' by completing the first two sections of a 'KWL' grid – things I **K**now, things I **W**ant to know, things I have **L**earned.

Challenge and develop

- As a class or in groups, the students **carry out investigations** to model the effects of the Earth's motion using Practical sheet 3.5.12. In a darkened room, first work with the students using a globe and a bright lamp or projector exploring or demonstrating the effect of the rotation of the Earth (activity 1 on the practical sheet). Ask them to **deduce** the time for one rotation (24 hours).
- They then **model** the Earth orbiting the Sun in four stages, improving the model each time (activity 2): [O1&2]
  - circular orbit with the Earth aligned vertically N to S with no rotation
  - elliptical orbit with the Earth aligned vertically N to S with no rotation
  - elliptical orbit with the Earth tilted with no rotation
  - elliptical orbit with the Earth tilted and rotating on its axis.

  Ask the students to **deduce** the time for one orbit (1 year), the number of rotations per orbit (365) and the tilt of the Earth in relation to the Sun at opposite sides of its orbit (North Pole pointed away/ towards). [O2]
- Using a globe on a stand and focusing on the section of the orbit where the northern hemisphere is tilted away from the Sun (the bright lamp or projector), stick a label on the UK. The students **observe** the effect of

the Earth's tilt on day length (shorter). Repeat the demonstration with the northern hemisphere tilted towards the Sun to show longer day length. Then contrast with the southern hemisphere. [O1&2]

- The students then **model** the effect of the Earth's tilt on the intensity of sunlight (energy) reaching the Earth (activity 3 on Practical sheet 3.5.12). [O1&2]
- Discuss as a class the questions on Practical sheet 3.5.12. [O1, 2&3]
- If possible, show the video 'The Seasons demo 1' (1 min 46 sec) from the Institute of Physics website – it illustrates why temperatures are lower in the winter. Go to www.iop.org and search for 'Teaching astronomy and space videos', then click on the first link in the list of results and scroll down to 'Classroom demonstrations'. [O1, 2&3]

Explain

- The students use Worksheet 3.5.12 or their own diagrams and annotations to **explain** how changes in day length can be accounted for by the tilt of the Earth's axis. [O1&2]

  Higher-attaining students may be able to **produce explanations** without prompts.

Consolidate and apply

- **Pairs to fours** The students **discuss** and **explain** how it is possible for polar regions to have 24-hour days in the summer and the opposite in the winter. [O1&2]
- They **discuss** and **explain** how days and seasons would be affected if the Earth's axis was tilted at 45° (compared to the current 23°) or tilted at 90° (45°: exaggerated variations; 90°: little variation). [O3]

Extend

- Ask students able to progress further to use the internet to **compare** average temperatures in the southern and northern hemispheres. (The southern hemisphere is colder due to the large area of ice and water, despite being closer to the Sun than the northern hemisphere is during the Earth's elliptical orbit.) [O2&3]

Plenary suggestions

Review Revisit the starter quiz, focusing on areas that students were not originally confident with. [O1, 2&3]

KWL grid If students started a KWL grid at the beginning of the lesson, they can now complete the 'L' section (things they have learned). [O1, 2&3]

Answers to Student Book questions

1. a) The Arctic Circle has no day; northern France has 7.5 hours. b) The Arctic Circle has a 24-hour day; northern France has 16.5 hours.
2. Day length equals night length (12 hours) in both locations. 3. The peak in day length happens on 21 Dec.
4. In the summer, the Earth is tilted towards the Sun.
5. Twice a year, around the spring and autumn equinoxes; because the tilt of the poles is not towards or away from the Sun.
6. In places on the equator the Sun never shines at an oblique angle, so seasonal changes are small. Day length also varies less than in places closer to the poles.
7. In the summer, the areas around the poles are always in sunlight, due to the tilt of the Earth on its axis.
8. Shorter days and sunlight at an oblique angle mean less energy for photosynthesis. Some may mention lower temperature, also reducing photosynthesis/growth.
9. The daily period is the same. There are more extreme differences in day and night than on Earth, because Mars is more tilted. Other seasonal variations are also more extreme due to the bigger tilt and more elliptical orbit (its distance from the Sun varies more). A year is twice as long on Mars as on Earth.

Answers to Worksheet 3.5.12

1. a) 24 hours b) no c) spherical d) The Earth is tilted away from the Sun. e) 12 hours each, with small variations
2. a) wide arc; but not going overhead b) shallower, narrower arc c) arc would pass overhead

   Note: it is hard to represent the arc in two dimensions; this could be discussed with students.
3. a) at A: total daylight; at B: constant dim light; at C: total darkness; at D: constant dim light

   b) at A and C: constant dim light; at B and D: even-length days and nights

   c) The poles would suffer cold winters of total darkness, and hot summers of total daylight. The equator would have winter at A and C, and summer at positions B and D, because the planet is nearer the Sun at positions B and D.

5.13 Measuring distances in the Universe

Lesson overview

Learning objectives

- Recall that the light year is used to measure astronomical distances.
- Explain the limitation of units such as kilometres in describing astronomical distances.
- Describe a technique for measuring the distance to distant objects.

Learning outcomes

- Explain the need for a large unit of astronomical distance. [O1]
- Describe how light years can be used to measure distance. [O2]
- Explain how the parallax method can be used to measure distance to some stars; explain the implications of an object being light years away. [O3]

Skills development

- Thinking scientifically: use units and nomenclature
- Working scientifically: develop explanations
- Learner development: communicate effectively

Resources needed mini-whiteboards; marker pens; A4 sheets of paper; pens; metre rulers; 5 m long strip of paper; graph paper; rulers; Worksheet 3.5.13; Practical sheet 3.5.13

Digital resources Quick starter; Interactive activity: Match the descriptions to the correct astronomical distances; Hangman: Key vocabulary game

Common misconceptions The light year is a measurement of time.

Key vocabulary light year (ly), parallax method

Teaching and learning

Engage

- **Mini-whiteboard activity** The students **estimate** the approximate distance from the Earth to the Sun (149 600 000 km or nearly 150 million km). They then try to **write out** the approximate distance to the next nearest star, Sirius – approximately 5 million times further away. [O2]

  Higher-attaining students can be introduced to scientific notation for writing extremely large numbers – for example 1 million = 1×10^6, 1 billion = 1×10^9.

Challenge and develop

- Ask the students to **list** other units and ideas they have encountered for measuring distances on Earth and in space (task 1 of Worksheet 3.5.13). [O1]
- Discuss the need to find a measure that is based on a standard distance. Ensure that the students **realise** that light always travels at the same speed in a vacuum (300 000 km/s). They **suggest** how they would work out how far light would travel in 1 year (by multiplying by the number of seconds in a minute, hour etc). [O1&2]

  Higher-attaining students can be asked to **convert** the speed of light into the distance travelled in longer time spans, such as an hour, day, week and year.

- Using the Student Book, introduce the light year as a measure of astronomical distance. Ask the students to **use units and nomenclature** to write the length of one light year (9 460 000 000 000 km or 9.46×10^{12} km) and also to write the figure in words (nine thousand, four hundred and sixty billion km – or nine trillion four hundred and sixty billion km). [O1&2]

- The students **compare** distances in the Universe. Ask them to read the 'Distances in the Universe' section in the Student Book and to answer questions 4 and 5. [O1&2]
- **Group work** In mixed ability groups of three, the students **investigate** how parallax causes the apparent position of an object to change by different amounts at different distances, using Practical sheet 3.5.13. [O3]
- Ask the students to **explain** why we see stars as they were in the past, rather than as they are now. (The light has taken a long time to reach us on Earth.) [O3]

Explain

- The students work in pairs to produce a 30 word **explanation** of why light years are useful for measuring distances in space. They then **communicate** their explanations to the class. [O1&2]
- They **discuss** and **explain** if and how it would be possible to travel to distant stars (task 2 of Worksheet 3.5.13). [O3]

Consolidate and apply

- Using a strip of paper 5 m long stretched across the room to represent the distance between the Earth and Sirius, the students estimate where they should mark the distance between the Earth and the Sun. (On this scale it would be 0.000 001 mm or one-millionth of a millimetre). [O1&2]
- Following on from their practical work on the parallax effect, students read 'Measuring techniques' in the Student Book and answer questions 6 and 7. [O1&3]

  Higher attaining students can complete task 3 of Worksheet 3.5.13. [O3]

Extend

- Ask students able to progress further to use the internet to **find out about** the brightness method for measuring distances from the Earth to a star. [O3]

Plenary suggestions

I think that is... because... The students suggest how feasible space travel is to distant stars. [O1, 2&3]

Answers to Student Book questions

1. kilometres per second 2.light years 3. The numbers are too big to be managed and compared easily.
4. a) Both have elliptical orbits (especially Pluto), and have different orbit times.
   b) The variation in distance is insignificant compared to the distances to other bodies.
5. The light from the more distant star has taken longer to reach us than light from the nearer star; therefore we see it further in the past. 6. Earth is at opposite sides of its orbit; so the difference in its view of the star is at its largest; this reduces the percentage error.
7. a) From Table 3.5.13: the ratio is 8.6 ÷ 0.000 016 = 537 500 times as far.
   b) 537 500 mm = 0.5(375) km

Answers to Worksheet 3.5.13

1. e.g. millimetre (mm); metre (m); kilometre (km); light year (ly)
2. a) Should show an appreciation of the need for the spaceship to be completely self-sustaining in all respects. Some high-attaining students may consider the massive timescale being too much for the survival of life.
   b) They may include ideas about the impracticality with current technology, scientific understanding and cost issues. They may identify expected and unexpected potential benefits from the project.
3. a)–d) They should draw accurate scale diagrams. Object A would appear to move by 60 mm (± 2 mm); object B would appear to move by 10 mm (±1 mm).
   e) The apparent movement is less, the further away the object is; so small errors make a big difference.

Answers to Practical sheet 3.5.13

1. It is more accurate for closer objects, because the apparent change in position of the object for a given change in observer position is greater.
2. Any suitable suggestion, e.g. the change of position of nearby objects such as trees against the background landscape when you move along in a car.

5.14 Checking students' progress

The *Checking your progress* section in the Student Book indicates the key ideas developed in this chapter and shows how students progress to more complex levels. It is provided to support students in:

- identifying the key ideas
- developing a sense of their current level of understanding
- developing a sense of what the next steps in their learning are.

It is designed either to be used at the end of a chapter to support an overall view of progress, or alternatively during the teaching of the chapter. Students can self assess or peer assess using this as a basis.

It would be helpful if students can be encouraged to provide evidence from their understanding or their notes to support their judgements. In some cases it may be useful to explore the difference in the descriptors for a particular idea so that students can see what makes for a 'higher outcome'.

It may be useful with some descriptors to provide examples from the specific work done, such as an experiment undertaken or an explanation developed and recorded. If marking and feedback use similar ideas and phrases, this will enable students to relate specific marking to a more general sense of progress.

To make good progress in understanding science students need to focus on these ideas and skills:

Students who are making modest progress will be able to:	Students who are making good progress will be able to:	Students who are making excellent progress will be able to:
Collect data about distance travelled and time taken for types of movement or journeys.	Present data you have collected or data you have been given as distance–time graphs.	Construct distance–time graphs for complex journeys.
Describe some features of distance–time graphs.	Analyse distance–time graphs to describe an object's movement at different stages in a journey.	Explain distance–time graphs for complex journeys, including where an object travels at different speeds and accelerates at different rates.
Describe a situation where objects are travelling at different speeds.	Apply the idea of relative speed to two objects moving in situations involving overtaking and collisions.	Apply the concept of relative motion to several moving objects in a variety of situations.
Identify the forces acting on an object and explain how they can cancel each other out so that a stationary object does not move.	Explain how opposing forces may or may not be in equilibrium and the effect that this has on a stationary object.	Explain how multiple forces may or may not be in equilibrium and identify the effect this has on an object.
Identify the forces acting on an object moving at a constant speed and explain how they can cancel each other out so that there is no change in speed or direction.	Explain how opposing forces may or may not be in equilibrium and the effect this has on an object's motion.	For a moving object, explain how multiple forces can act and how they may or may not be in equilibrium; identify the effect this has on an object's motion.
Explain what causes an object to have weight.	Describe how gravity affects the weight of an object.	Explain the relationship between gravitational field and the weight of an object.
Describe how an object's weight can vary.	Predict how an object's weight would vary depending on its position in relation to large bodies such as planets.	Use the concept of gravitational field to explain various phenomena, including the orbits of planets around stars.
Describe the movement of the Sun, Earth and Moon in relation to each other.	Explain the effects of the relative motion of the Sun, Earth and Moon.	Explain the relative movement of the Sun, Earth and Moon using the idea of gravity.
Describe the differences between the Sun, other stars and galaxies.	Describe the relationship between the Sun, other stars and galaxies.	Relate ideas about the Sun, stars and galaxies to evidence visible from the Earth.
Describe the effects that the tilt of the Earth's axis has on Earth.	Explain the causes of daily and seasonal changes.	Explain what would happen if the Earth's axis was tilted by a different amount.
Recognise the need for large units to measure distances in space.	Describe how light years can be used to measure distances.	Explain the implications of other stars and galaxies being many light years away from Earth.

5.15 Answers to Student Book Questions

This table provides answers to the Questions section at the end of Chapter 5 of the Student Book. It also shows how different questions assess attainment in terms of the focus and style of a question as well as the context. Question level analysis can indicate students' proficiency in approaching different aspects of scientific understanding and different types of answer.

Q	Answer		Focus			Style			Context			
		Marks available	Knowledge & understanding	Application	Evaluation of evidence	Objective test question	Short written answer	Longer written answer	Motion and equilibrium	Gravitational fields	The Earth and the Solar System	Stars and galaxies
1	c	1	x			x			x			
2	d	1	x			x						x
3	a	1	x			x				x		
4	d	1	x			x			x			
5	• Light years (ly)	1	x				x					x
	• The distances are so big, it makes the numbers easier to deal with	1	x				x					x
6	• The Earth rotates on its axis once every 24 hours	1	x				x				x	
	• On the side of the Earth facing the Sun it is day; and on the opposite side it is night	1	x				x				x	
7	• Short days compared to nights in winter; opposite in summer	1	x					x			x	
	• Lower temperatures in winter compared to summer											
	• The Earth is tilted on its axis	1	x					x			x	
	• In winter the northern hemisphere spends more time every 24 hours tilted away from the Sun; the Sun's energy arrives at a shallow angle so there is less heat	1 1	x x					x x			x x	
8	a	1		x		x				x		
9	c	1		x		x			x			
10	It would carry on in a straight line at a tangent to the orbit	1		x		x			x			
11	d	1		x		x				x		

Q	Answer		Focus			Style			Context			
		Marks available	Knowledge & understanding	Application	Evaluation of evidence	Objective test question	Short written answer	Longer written answer	Motion and equilibrium	Gravitational fields	The Earth and the Solar System	Stars and galaxies
12	• 10 N (resultant) pushing force	1		x			x		x			
	• The scooter will get faster/ accelerate	1		x			x		x			
13	Accept any **two** from:											
	• More of the northern hemisphere would have 24-hour darkness in the winter; opposite in the summer	2		x			x				x	
	• Summers would be hotter because the Sun is more directly overhead											
	• The North Pole would have a rapid transition from 24-hour nights to 24-hour days											
14	• Distance–time graph with *x*-axis labelled 'Time' over 0– 4 minutes; *y*-axis labelled 'Distance' over 0–500 metres	1		x				x	x			
	• for the first horse, a straight line from point (0, 0) to point (4, 500)	1		x				x	x			
	• for the second horse, initially an upward curve from point (0, 0)	1		x				x	x			
	• then the line is horizontal for 1 minute , then an upward curve finishing at point (3, 500)	1		x				x	x			
15	a) iii	1			x	x				x		
	b) … the motion of the ball is fairly slow so air resistance is limiting the ball's movement by only a small amount	1			x	x			x			
16	• Objects, including stars, orbiting around the same point suggests something with a very strong gravitational field	1			x			x		x		
	• A dark area of space at the location of the black hole because no light can escape the strong gravitational field	1			x			x		x		
	• As the spaceship entered the gravitational field of the back hole, it would accelerate towards the black hole	1			x			x		x		
	• If it got too close its engines would not be able to overcome the gravitational attraction	1			x			x		x		
	Total possible:	**30**	**12**	**12**	**6**	**10**	**8**	**12**	**11**	**8**	**8**	**3**

6.1 Introduction

When and how to use these pages

The Introduction in the Student Book indicates some of the ideas and skills in this topic area that students will already have met from KS2 or from previous KS3 work, and provides an indication of what they will be studying in this chapter. *Ideas you have met before* is not intended to comprehensively summarise of all the prior ideas, but rather to point out a few of the key ones and to support the view that scientific understanding is progressive. Even though students might be meeting contexts that are new to them, they can often use existing ideas to start to make sense of them.

In this chapter you will find out indicates some of the new ideas that the chapter will introduce. Again, it isn't a detailed summary of content or even an index page. Its purpose is more to act as a 'trailer' and generate some interest.

The outcomes, then, will be recognition of prior learning that can be built on, and interest in finding out more.

There are a number of ways this can be used. You might, for example:

- Use *Ideas you have met before* as the basis for a revision lesson as you start the first new topic.
- Use *Ideas you have met before* as the centre of spider diagrams, to which students can add examples, experiments they might have done previously or what they found interesting.
- Make a note of any unfamiliar/difficult terms and return to these in the relevant lessons.
- Use *In this chapter you will find out* to ask students questions such as:
  - Why is this important?
  - How could it be used?
  - What might we be doing in this topic?

Overview of the chapter

In this chapter, students will learn about and compare the properties of waves in water and light waves and how they transfer energy. The students will also learn about the reflection, refraction and absorption of light by materials. They will learn that light consists of tranverse waves with a range of frequencies and wavelengths, and that white light can be split into a spectrum of colours.

The students will learn that energy can be stored in various ways and that it is transferred from one form of energy store to another when a change occurs (either physical or chemical). They will also learn that quantities of stored energy and transferred energy can be measured and/or calculated, and learn the correct use of units. The students will investigate fuels and food as energy stores, and learn about fuel bills and the use of energy in homes. They will learn about the importance of the rate of energy transfer.

This chapter offers the opportunity for students to use and evaluate models to explain the properties of light in terms of waves. They will also analyse and evaluate primary and secondary data linked to energy stores and transfers.

Obstacles to learning

The students may need extra guidance with the following terms and concepts:

- **Waves** A mechanical wave moves through a material, but there is no net movement of that material – for example in water a wave passes through it, but the water simply oscillates up and down. Light waves (electromagnetic waves) do not need a medium – they can travel through a vacuum.
- **Vacuum (empty space)** Some students may still think of air as 'empty space', yet sound is a mechanical wave. They need to understand that an open 'empty' container is filled with air particles, so it is not really empty. To produce a vacuum (empty space) all the air must be sucked out of a container.
- **Reflection and absorption** Some students may think that an object either absorbs or reflects light, and that it cannot do both. However, all objects absorb and reflect light to different degrees.
- **Energy transfer and change** It is often thought that the transfer of energy causes change. It does not – energy is transferred as a result of change. Also, energy is not 'used up' or 'produced'. The quantity of energy

stored before the change is the same as the quantity stored after the change, even though the form of storage has changed (conservation of energy).

	Topic title:	**Overarching objectives:**
2	Making waves	Waves on water as undulations which travel through water with transverse motion; these waves can be reflected, and add or cancel – superposition
3	Exploring light waves	The similarities and differences between light and waves in matter; light waves travelling through a vacuum; speed of light
4	Explaining properties of light waves	The transmission of light through materials: absorption, diffuse scattering and specular reflection at a surface
5	Using the ray model	Use of the ray model to explain imaging in mirrors, the pinhole camera, the refraction of light and action of convex lens in focusing (qualitative); the human eye
6	Understanding energy transfer by light	Light transferring energy from source to absorber, leading to chemical and electrical effects; photosensitive material in the retina and in cameras
7	Exploring coloured light	Colour and the different frequencies of light, white light and prisms (qualitative only); differential colour effects in absorption and diffuse reflection
9	Understanding energy transfer and change	Energy as a quantity that can be quantified and calculated; the total energy has the same value before and after a change Comparing the starting with the final conditions of a system and describing increases and decreases in the amounts of energy associated with movements, temperatures, changes in positions in a field, in elastic distortions and in chemical compositions Using physical processes and mechanisms, rather than energy, to explain the intermediate steps that bring about such changes
10	Explaining thermal conduction and radiation	Heating and thermal equilibrium: temperature difference between two objects leading to energy transfer from the hotter to the cooler one, through contact (conduction) or radiation; such transfers tending to reduce the temperature difference: use of insulators
11	Understanding energy transfers by fuels and food	Energy as a quantity that can be quantified and calculated; the total energy has the same value before and after a change Comparing energy values of different foods (from labels) (kJ) Other processes that involve energy transfer: metabolism of food, burning fuels
12	Comparing rates of energy transfer	Comparing power ratings of appliances in watts (W, kW) Comparing amounts of energy transferred (J, kJ, kW hour)
13	Looking at the cost of energy use in the home	Comparing power ratings of appliances in watts (W, kW) Comparing amounts of energy transferred (J, kJ, kW hour) Domestic fuel bills: fuel use and costs

6.2 Making waves

Lesson overview

Learning objectives

- Describe the movement of waves in water.
- Understand reflection of waves.
- Understand superposition of waves.

Learning outcomes

- Describe how ripples and waves move in water. [O1]
- Explain that waves can be reflected. [O2]
- Explain how waves can add or cancel out. [O3]

Skills development

- Thinking scientifically: ask questions
- Working scientifically: design investigations
- Learner development: communicate effectively

Resources needed a long rope (to model how transverse waves travel); materials and equipment as detailed in the Technician's notes; Worksheet 3.6.2; Practical sheet 3.6.2; Technician's notes 3.6.2

Digital resources Quick starter; Slideshow: Making waves - A look at the up and down movement of waves; Video; Interactive activity: Define the key terms about waves and wave motion

Common misconceptions Waves and ripples carry water in the direction in which they move.

Key vocabulary reflection, longitudinal wave, transverse wave, oscillate, superposition

Teaching and learning

Engage

- Show the students some photographs or a video of waves in the sea. Working in small groups, ask them to **discuss** how water moves. After a few minutes invite each group to **summarise** its ideas. [O1]

Challenge and develop

- Following the class discussion, ask the students to answer questions 1–3 in the Student Book and complete task 1 of Worksheet 3.6.2. [O1]
- **Pairs/small groups** Ask the students to **carry out** the first two experiments on Practical sheet 3.6.2, investigating ripples in a tray of shallow water. It may be helpful to first demonstrate how to produce a single ripple or series of ripples. Ask the students to work carefully, not agitating the water so much that it spills. The students should **make notes** of their observations. [O1&2]
- Ask each pair or group to make a brief **presentation** to the class, outlining their observations. [O1&2]

Explain

- To reinforce the idea that water moves up and down in a ripple or wave, ask the students to **carry out** experiment 3 of Practical 3.6.2. They should **make notes** of their observations. [O1&2]
- Explore with the students how the results from Practical 3.6.2 support the idea that water goes up and down but does not move in the direction of the wave. [O1]
- Invite two students to **demonstrate** making waves in a length of rope. One student should try to hold one end of the rope so that it does not move. The other holds the other end of the rope, pulling it so that the rope is only slightly loose. This second student then sends pulses down the rope by jerking the end. Ask others in the class how this is similar to waves in water. Explain that waves in water and waves in a piece of rope are examples of mechanical waves (an idea that will be met again later). [O1]

- Now ask all the students to complete task 2 of Worksheet 3.6.2 and to answer questions 4–6 in the Student Book. [O2]

Consolidate and apply

- Ask the students to look at the 'Superposition' section in the Student Book. Discuss the idea of superposition and ask the students to **sketch diagrams** to show the situation when the positions of the two waves are somewhere between the two extremes (crests on one overlapping with crests on the other; crests on one overlapping with troughs on the other). [O3]
- Ask the students to answer questions 7 and 8 in the Student Book. Ask some students to **share** their answers to check their understanding. [O3]
- Review the evidence that in transverse mechanical waves – like in water or on a rope – the particles in the material that carries the wave move up and down, but do not move in the direction of the wave. This leads naturally to the suggested plenary. [O1&2]

Extend

- Ask students able to progress further to do task 3 of Worksheet 3.6.2. This will consolidate their observations that the shapes of waves can be changed and will introduce them to ideas of wavelength and amplitude, which they should be prompted to remember from their work on sound waves in Year 7. [O1&2]

Plenary suggestions

- **Mexican wave** Arrange the students in a line. Ask them to do a Mexican wave by moving from the crouching position (a trough) to standing up with their arms raised high (a crest). Each student in the line should carry out the movement a second or so after the student before them. Explain that the overall movement is a transverse wave – students only move up and down, but the wave (energy) moves forward.

Answers to Student Book questions

1. the light has bounced back
2. it comes back at the same angle as it hit the barrier
3. A wave bounces back in different directions, depending on what part of the surface it hits.
4. water oscillates up and down as the wave moves through it
5. a crest is the highest point of the oscillation and a trough is the lowest point
6. Compared with the direction of the wave, transverse waves oscillate up and down, and longitudinal waves oscillate forwards and backwards.
7. When two waves meet, they add to one another.
8. The two waves cancel one another out, so there is no wave.

Answers to Worksheet 3.6.2

1. splash; circles; bobs; away
2. a) and b) should both be typical transverse waves, as shown in the Student Book.
   c) because the material carrying the wave goes up and down at right angles to the direction in which the wave moves
   d) crests are the highest points; troughs are the lowest points
3. a) a sketch of a regular wave of longer wavelength than that on the worksheet, with one wavelength labelled
   b) a sketch of a regular wave of higher amplitude than that on the worksheet, with the amplitude labelled
   c) suitable experiments using the tray of water and a paddle; by varying the force and the depth of the paddle movements generating the waves

6.3 Exploring light waves

Lesson overview

Learning objectives

- Describe light as travelling in waves.
- Understand the similarities and differences between water waves and light waves.
- Explain the frequency of a wave.

Learning outcomes

- Recall that light travels in waves. [O1]
- Compare the properties of water waves and light waves. [O2]
- Explain what is meant by the frequency of a wave. [O3]

Skills development

- Thinking scientifically: ask questions
- Working scientifically: develop explanations
- Learner development: collaborate effectively

Resources powerful torch (e.g. Maglite); length of rope; electric bell in a jar attached to a vacuum pump; Worksheet 3.6.3

Digital resources Quick starter; Interactive activity: Drag the statements about light and sound into the correct group - true or false

Common misconceptions Light goes around things, not just in straight lines. Air is empty space.

Key vocabulary crest, trough, frequency, hertz, wavelength

Teaching and learning

Engage

- **Pairs/small groups** Ask the students to **discuss** and **explain** why it is light during the day, but dark at night. Draw out the idea that light comes from the Sun. They are likely to know that light travels in a straight line; explain that the next question to answer is 'How does it get to Earth?'. [O1]

Challenge and develop

- Shine a torch around a dimmed, but not dark, classroom. Ask the students to **share** what evidence they think there is that light travels in straight lines. [O1]
- Ask the students to answer questions 1 and 2 in the Student Book and to complete task 1 of Worksheet 3.6.3. Use this to reinforce their understanding of light sources and reflected light. [O1]

Explain

- Remind the students how waves move in water – they travel in one direction as the water oscillates up and down at right angles to that direction – and how waves are reflected. Draw comparisons with light and introduce the idea that light also travels as transverse waves. Ask the students to answer questions 3–5 in the Student Book. [O1&2]
- Explain that light also travels as transverse waves but unlike waves in water, light waves do not need a medium to travel through. They can move through a vacuum (empty space). If possible, demonstrate the experiment summarised in Figure 3.6.3d of the Student Book. [O1&2]
- Review the idea of wavelength. This can be illustrated by making mechanical waves in a length of rope or ripples in water. [O3]

Consolidate and apply

- Ask the students to complete task 2 of Worksheet 3.6.3. [O2]
- Having compared water and light waves, ask the students to **describe** the movement when a length of rope anchored at one end is jerked to produce a transverse wave, using what they have learned so far about waves. [O2&3]
- Ask the students to answer questions 6 and 7 in the Student Book. [O3]

  Higher-attaining students can also attempt Student Book question 8 and task 3 of Worksheet 3.6.3.

Extend

- Ask students able to progress further to **think about** how fast light travels. Tell them that it takes about 8 minutes and 20 seconds for light from the Sun to reach Earth. Ask them what other information they need to calculate the speed of light – they need the distance, which is about 150 000 000 km. [O3]

Plenary suggestions

Travelling light Arrange the students in small groups. Ask them to **present** a short, illustrated description of how light travels through space. After the class has heard all the presentations, ask them to **decide** and **vote** for the one they think was the clearest and most accurate. [O1, 2&3]

Answers to Student Book questions

1. suitable drawings showing: a) a light wave travelling straight from the flame to the eye; b) a light wave from a source being reflected from a person's face to another person's eyes
2. less light is available to be reflected
3. for example: a spring stretched or squashed and then released; a rope held at both ends and being shaken; sound waves
4. it oscillates at right angles to its direction of travel
5. it produces sound waves; these are mechanical waves that cannot travel through empty space
6. A frequency is the number of waves that pass a fixed point each second.
7. The higher the hertz value, the higher the wave frequency, or number of waves that pass a point every second.
8. the wavelength will be shorter

Answers to Worksheet 3.6.3

1. a) the display is a light source that operates only when energy is transferred from an energy store (mains electricity or battery)

   b) the lit-up screen is a source of light. The eye sees other parts of the tablet because they reflect light from the Sun or other light sources into the eye
2. a) light waves

   b) Light is an electromagnetic wave that does not need a medium to travel through. The oscillations are up and down at right angles to the direction in which the wave travels.
3. a) Students need to identify two crests (or troughs), use a ruler and measure the distance between them.

   A = 2.0 cm; B = 0.6 cm; C = 1.2 cm

   b) wave B has the highest frequency; wave A has the lowest frequency

6.4 Explaining properties of light waves

Lesson overview

Learning objectives

- Describe how light passes through different materials.
- Understand how light can be absorbed by materials.
- Explain the difference between diffuse scattering and specular reflection.

Learning outcomes

- Recall that light passes through transparent materials. [O1]
- Explain how light is absorbed by opaque materials. [O2]
- Compare diffuse scattering and specular reflection. [O3]

Skills development

- Thinking scientifically: consider the quality of evidence
- Working scientifically: record evidence
- Learner development: ask questions

Resources samples of transparent, translucent and opaque materials, including frosted glass; blank cards for the loop game; materials and equipment as detailed in the Technician's notes; Worksheet 3.6.4; Practical sheet 3.6.4; Technician's notes 3.6.4

Digital resources Quick starter; Slideshow: Windows, shadows and mirrors; Interactivity: Order the materials, from the most transparent to the most opaque

Common misconceptions Light can only be reflected from shiny surfaces (such as a mirror). An object either absorbs or reflects light, and cannot do both.

Key vocabulary transparent, opaque, translucent, specular reflection, image, diffuse scattering, absorb

Teaching and learning

Engage

- Review the idea of waves by asking the students to complete task 1 of Worksheet 3.6.4. They should come to the **conclusion** that light travels much faster than sound. Raise the question 'But does light travel at the same speed through all materials?' [O1]

Challenge and develop

- Show the class samples of materials that are transparent, translucent and opaque – but don't mention the terms at this stage. Organise the students into pairs or small groups and ask them to **sort** the materials into groups and to **explain** their reasons for the groupings they choose. Ask them to answer questions 1 and 2 in the Student Book. [O1]
- In their pairs, ask the students to **carry out an investigation** into transparent, translucent and opaque plastic materials, following the instructions on Practical sheet 3.6.4. Remind them that they should **make notes** of their observations. [O2]

Explain

- When the students have completed the practical investigations, ask them to **share** their results. [O2]
- Through discussion introduce the idea of light passing through transparent materials and being reflected by opaque materials. Introduce the term 'specular reflection' to describe reflection at smooth, shiny surfaces. Explain that this is why they can see reflections of themselves when they look in, for example, a mirror or a highly polished car. [O3]

- Ask the students to **suggest** what happens when light strikes a translucent material. Introduce the idea of 'diffuse reflection'. Ask them to look at Figure 3.6.4e in the Student Book to understand how diffuse reflection happens. [O3]
- Ask the students to answer questions 4–6 in the Student Book and to complete task 2 of Worksheet 3.6.4. [O3]

Consolidate and apply

- **Pairs/small groups** Give the students some transparent plastic cups (from Practical 3.6.4). Ask them to breathe on the cups and to note what happens – the surface of the cup becomes fogged up. Invite them to **explain** what is happening. Guide them to the idea that the 'fog' is water from their breath that has condensed on the cup. It consists of tiny droplets – these produce diffuse reflection. [O2&3]
- If available, get a piece of frosted glass. Ask whether it is transparent, translucent or opaque, and then what happens when light strikes it. Now stick a piece of clear sticky tape to the surface. Ask the students what has happened and why? The area with the tape becomes transparent – this is because the tape has a smooth, flat surface that produces specular reflection. [O2&3]

Extend

Ask students able to progress further:

- to **explain** why polishing things makes them shine. [O3]
- to read the 'Absorption of light' section of the Student Book, answer questions 7 and 8, and then to complete task 3 of Worksheet 3.6.4. [O2&3]

Plenary suggestions

- **Loop game** Ask each student to **write a question** on a piece of card about how light is absorbed by different materials, and then to write (on a different-coloured card) the answer. Collect, shuffle and redistribute the cards so that each student has one question and one answer. The first student reads out the question they have and the student with the correct answer answers the question. This second student then reads out what is written on their question card. Continue the game until all the questions have been answered.

Answers to Student Book questions

1. for example: water; glass; clear polythene; air
2. Shadows from opaque objects are darker than those from translucent objects.
3. it blocks all the light from passing through
4. *specular reflection*: all the light bounces back at the same angle; *diffuse reflection*: light bounces back in various random directions
5. A rough surface has lots of small, flat areas at different angles to one another; the light bounces back in many different directions.
6. The light is reflected in one direction when the water is calm (flat); it is reflected back in various directions when the water is choppy.
7. light passes through the ice and is transmitted from the opposite side
8. light is reflected back from the flat, shiny surface

Answers to Worksheet 3.6.4

1. Light travels the fastest; sound is next; the ripples of water are slowest.
2. reflected; direction; specular; various; diffuse
3. a) A perfectly smooth, flat metal surface reflects light (specular reflection). However, it may scratch fairly easily or become tarnished or corroded. The glass protects the metal surface.

   b) Particles in glass and in metal have different properties. Light passes through the transparent glass layer (a small amount is reflected). All the light that reaches the metal layer is reflected.

Answers to Practical sheet 3.6.4

Results: A is transparent; B is translucent; C is opaque.

Analysis: The observations suggest that light waves strike transparent materials and pass straight through. When light waves strike translucent materials, some pass straight through, while some are reflected. When light waves strike opaque materials no light passes through, it is all reflected.

6.5 Using the ray model

Lesson overview

Learning objectives

- Describe the ray model of light.
- Explain how the direction of light rays can be changed.
- Explain how a pinhole camera and the eye work.

Learning outcomes

- Recall that the ray model is a way of showing the direction of light and how it changes. [O1]
- Explain what happens when light is reflected and when it is refracted. [O2]
- Draw ray diagrams to show how the eye works. [O3]

Skill development

- Thinking scientifically: analyse data
- Working scientifically: interpret evidence
- Learner development: communicate effectively

Resources materials and equipment as detailed in the Technician's notes; drawings of pinhole camera and the eye (if models aren't available); protractors; Worksheet 3.6.5 (the final page copied onto card); Practical sheet 3.6.5; Technician's notes 3.6.5

Digital resources Quick starter; Interactive activity: Reorder the sentences to describe how light enters the eye

Common misconceptions Light goes around things, not just in straight lines.

Key vocabulary ray model, refraction, cornea, lens, retina

Teaching and learning

Engage

- **Small groups** Hand out a plane mirror to each group of students, and possibly a torch (in which case darken the room) – but warn students not to look directly at the beam or its reflection if the torches are bright. Ask the students to experiment with the mirror and to use their observations to **draw conclusions** about the way light travels and how it is reflected. [O1&2]

Challenge and develop

- As a whole class exercise, encourage the students to **discuss** what they have learned from experimenting with plane mirrors. [O1&2]
- Summarise the main features of the ray model of light. Ask the students to look at Figure 3.6.5b in the Student Book, discuss the angle of reflection and the conventions used in such diagrams, and then ask them to work individually to complete task 1 of Worksheet 3.6.5. [O1&2]
- Ask the students to answer questions 1–5 in the Student Book. [O1&2]

Explain

- Invite some students to **show** their completed diagrams from the first task of Worksheet 3.6.5 to the class and to **explain** how they decided what to draw. [O2]
- Ask the students to complete task 2 of Worksheet 3.6.5. [O2]
- Ask the students to read 'The human eye' section in the Students Book and to answer questions 6–8. It would be helpful to show the students a model of an eye, if one is available. [O3]

Consolidate and apply

- Demonstrate the effects of refraction using a large glass jar of water and a metre ruler (or similar). Put the ruler in the water at an angle to show how it appears to 'bend'. Then put it in upright and move it slowly to one side of the beaker, back again and then to the other side and back again. Ask the students to **describe** what they see. [O1&2]
- **Pairs/small groups** Ask the students to **carry out an investigation** of apparent depth following the instructions on Practical sheet 3.6.5. [O1&2]
- If possible, show the students a pinhole camera (otherwise describe one). Ask the students to **compare** the pinhole camera with a human eye. Then ask them to complete task 3 of Worksheet 3.6.5. [O2&3]

Extend

- Ask students able to progress further the question 'Why do some people need to wear glasses?'. [O2]

Plenary suggestions

Differences and similarities Put a pinhole camera and a model of the eye side-by-side for students to **study**. If these are not available, provide two drawings that have no light rays included. Arrange the students in groups of four and ask them to **identify** differences and similarities between the camera and the eye. Ask each group to **present** its ideas to the class. After the presentations, ask the the class to **peer-assess** and vote for which presentation was the clearest and most accurate.

Answers to Student Book questions

1. Light rays from the source are reflected by the mirror into a person's eye. The light appears to be coming from behind the mirror.
2. a) longer lines are needed if the object is further away; the image appears further back in the mirror
   b) the opposite is true when objects are closer to the mirror
3. they slow down and are refracted (change direction)
4.

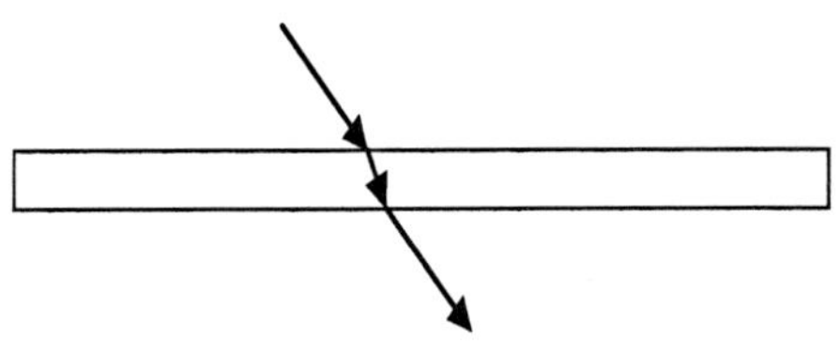

5 cornea; lens
6. it refracts light, so that rays emerging from the lens are focused on a common point
7. Light rays reflected from the top and bottom of the object are refracted. They pass through a common point and when they reach the retina, the rays that were originally reflected from the top of the object are at the bottom of the image.

Answers to Worksheet 3.6.5

1. correct ray diagrams, similar to Figure 3.6.5b in the Student Book; angles of incidence and reflection should be equal; extended dashed lines behind the mirror should meet at the image position; image should be same distance behind mirror as object is in front
2. Light travels through transparent materials, such as glass and air.
3. a) the light rays cross over where they pass through the pinhole, so the ray from the top of the object forms the bottom of the image
   b) correct ray diagrams, similar to that on the worksheet i) The image is smaller if the object is further from the pinhole. ii) The image is larger if the object is nearer to the pinhole. iii) The image is larger if the object is larger.
   c) i) The size of the image is inversely proportional to the distance of the object from the pinhole.
   ii) The size of the image is proportional to the size of the object.

Answers to Practical sheet 3.6.5

Results: The ratio of the real depth to the apparent depth should be about 1.33 (the refractive index of water).

Analysis:

1. Refraction is the slowing down (or speeding up) of a wave on entering a different medium.
2. Refraction of light on entering water causes light rays to change direction. This results in an image being seen, above the object.

6.6 Understanding energy transfer by light

Lesson overview

Learning objectives

- Describe light as a way of transferring energy.
- Give examples of chemical and electrical effects when materials absorb light.
- Explain changes that happen when materials absorb light.

Learning outcomes

- Recall that light transfers energy from place to place. [O1]
- Describe examples of chemical and electrical effects caused when materials absorb light. [O2]
- Explain in outline photosynthesis, the photoelectric effect and photochemical smog. [O3]

Skills development

- Thinking scientifically: analyse data
- Working scientifically: record evidence
- Learner development: collaborate effectively

Resources video of a solar car; equipment and materials as detailed in the Technician's notes; Worksheet 3.6.6; Practical sheet 3.6.6; Technician's notes 3.6.6

Digital resources Quick starter; Interactive activity: Drag the objects into the correct group - source of light, or reflector of light; Slideshow: Making use of light energy; Video

Key vocabulary energy transfer, photosynthesis, photochemical reaction, photovoltaic effect

Teaching and learning

Engage

- Show the class a video of a solar car – there are many examples on the internet. Discuss with the students why these cars move when they are in sunlight. [O1]

Challenge and develop

- Remind the students of the terms 'reflect', 'refract' and 'absorb', in the context of light striking different materials. Ask them to complete task 1 of Worksheet 3.6.6 and to answer questions 1–3 in the Student Book. [O1]
- **Pairs/small groups** Ask the students to **carry out** experiment 1 described on Practical sheet 3.6.6. This involves investigating how electrical devices can be powered by a solar cell. Ask the students to **record** all their observations. [O2]

Explain

- Discuss the observations made when the students carried out the practical work. Draw out the idea that the function of the electrical components depends on how much of the photovoltaic cell is exposed to sunlight and, therefore, the amount of energy that is transferred. [O1&2]
- Ask the students to read the 'Photosynthesis and photovoltaic cells' section in the Student Book and to answer questions 4–6. [O2&3]

Consolidate and apply

- Ask the students to complete task 2 of Worksheet 3.6.6. [O2]

  You may need to help some students.

- Following on from the students' discussions about their observations from the first part of Practical 3.6.6, ask them to **carry out** experiment 2 of Practical sheet 3.6.6. Invite students to **share** their results with the class and to **describe the relationship** between voltage and the exposed area of the photovoltaic cell. [O2&3]
- Ask the students to complete task 3 of Worksheet 3.6.6 and to answer Student Book questions 7 and 8, about the formation of photochemical smog. [O3]

Extend

- Ask students able to progress further to work together to **research** the use of solar panels on houses and other buildings, addressing issues such as how much voltage can be produced, what surface area of photovoltaic cells is needed and what happens when it is a cloudy day or night time. [O2]

Plenary suggestions

Sharing understanding Ask the students to **write two lists** about what they have been learning in the lesson:

- things I'm confident I understand
- things I'm not sure about or do not understand.

Invite them to **share** which parts of the lesson they did not understand. Ask those who are confident that they do understand to **explain** them. [O1, 2&3]

Answers to Student Book questions

1. Light from a light source enters the eye directly. Reflected light is light originally from a light source, but that has been reflected by something and travelled on into the eye.
2. it is transferred to particles of the material
3. a chemical change may happen; or an electric current may be produced; or it may just warm up slightly
4. Energy carried by light waves is transferred to particles in a material, resulting in a flow of electric current.
5. highly purified silicon
6. One cell produces only a small voltage, so large numbers are needed. It is the same reason that batteries consist of several cells connected in series.
7. Chemicals that cause pollution can absorb energy carried by light waves and produce a photochemical smog.
8. Light waves are absorbed and nitrogen dioxide particles (molecules) vibrate faster; if they vibrate fast enough, some oxygen atoms break away from molecules.

Answers to Worksheet 3.6.6

1. a) i) bounced back (reflected light) ii) passes through (transmitted) with its direction changed iii) energy transferred to the material, neither reflected nor transmitted

   b) When light waves are absorbed, energy is transferred to particles of the material, increasing their reactivity or motion. In reflection, the energy is bounced back and transferred away from the surface. In refraction, the energy is transmitted through the material.
2. A photovoltaic material absorbs sunlight, which releases electrons from the material's particles.
3. a)

A chlorine molecule, Cl_2 — Two chlorine atoms, Cal + Cal

   b) The chlorine molecules absorb light and vibrate faster. Eventually the energy transferred to the molecules is bigger than the energy that holds the two chlorine atoms in a molecule together; chlorine molecules split into two atoms.

Answers to Practical sheet 3.6.6

Results and conclusions: Components only function when the PV cell is exposed to light and produces an electric current. The more surface area exposed the brighter the lamp, the faster the motor and the louder the buzzer. The voltage produced should be directly proportional to the area exposed.

6.7 Exploring coloured light

Lesson overview

Learning objectives

- Describe how a spectrum can be produced from white light.
- Compare the properties of light of different frequencies.
- Explain how light of different wavelengths can be split and recombined.

Learning outcomes

- Recall ways that a spectrum can be made, including using a prism. [O1]
- Explain that the colour of light in a spectrum depends on its frequency. [O2]
- Explain that the higher the frequency, the shorter the wavelength and the more light is refracted. [O3]

Skills development

- Thinking scientifically: analyse data
- Working scientifically: develop explanations
- Learner development: respect others

Resources equipment and materials as detailed in the Technician's notes; Worksheet 3.6.7 (printed in colour); Practical sheet 3.6.7; Technician's notes 3.6.7

Digital resources Quick starter; Slideshow: Separating and combining colours; Interactive activity: Order the colours of the spectrum, from the shortest wavelength to the longest; Slideshow: Explaining refraction; Hangman: Key vocabulary game

Common misconceptions White light is made up of discrete colours. Colour is a property of objects rather than of reflected light.

Key vocabulary wavelength, spectrum, frequency

Teaching and learning

Engage

- **Small groups** Arrange the students in groups of three and ask them to **carry out** the practical work described in Practical sheet 3.6.7. This involves investigating what happens when white light passes through different-coloured pieces of transparent plastic. This is best done either outside or close to a window on a sunny day. Artificial light sources, such as torches, do not produce good spectra. Groups of three work well – one student holding the mirror, one the white card and the other the transparent plastic sheets. [O1]

Challenge and develop

- Ask the students to **describe** ways that spectra can be produced, other than by the method investigated in the practical. If possible, demonstrate prisms, diffraction gratings and a fine water spray. [O1]
- Ask the students to **think about** what happens when the coloured transparent sheets are used, prompting them if necessary by mentioning absorption and transmitted light. [O1]

Explain

- Talk about the visible spectrum, emphasising that the spectrum is continuous – has no sharp boundaries. Ask the students to complete task 1 of Worksheet 3.6.7 and to answer questions 1–3 in the Student Book. [O1]
- Ask the students to read the 'Different wavelengths' section in the Student Book, and to answer questions 4–6. They may do these individually or in small groups or through a whole class discussion. It may be necessary to spend a little time ensuring that the students have grasped the idea that the extent of diffraction depends on the wavelength of the light wave. [O2&3]
- Ask the students to work individually to complete task 2 of Worksheet 3.6.7. When they have made their measurements, suggest that they work in small groups to **discuss** the 'scale model' explanation. The

diagram of three light waves on Worksheet 3.6.7 is drawn so that the wavelengths are in a similar ratio to those of 'typical' waves in the red, green and blue regions of the visible spectrum. [O2&3]

Some students are likely to need guidance and prompting.

Consolidate and apply

- Ask the students to **recall** what they know about the ray model of light. Check that they remember drawing diagrams to show how light rays are reflected and refracted. [O3]
- **Pair talk** Ask the students to work individually to complete task 3 of Worksheet 3.6.7. Then give them some time to **compare** their own completed diagrams with those of a partner. If there are differences, encourage **discussion** about what they have drawn in their pairs and, if it helps, with another pair. [O2&3]

Extend

- Ask students able to progress further to **look again** at the way a spectrum was produced using a mirror and water. From their understanding of using a prism to make a spectrum, and the idea of diffraction being dependent on wavelength, ask them to **draw ray diagrams** to **explain** the spectrum that is produced using a mirror and water. [O2&3]

Plenary suggestions

Loop game Arrange the students in groups of five. Each student **writes a question** on a card about the visible spectrum and the wavelengths of light of different colours. They write the answer to their question on a different-coloured card. Collect and redistrubute the cards so that each student has one question and one answer. Now, ask one student to read out the question they are holding – and the student with the correct answer card has to identify themselves and read out the answer. This student then reads the question on the card they were given, and so on. [O1, 2&3]

Answers to Student Book questions

1. it is a triangular block (like a block of Toblerone)
2. it is a continuous spectrum of colours that changes smoothly from red to violet
3. red; violet
4. The shorter the wavelength, the higher the frequency.
5. red light
6. White light consists of waves with differing wavelengths. The shorter its wavelength, the more a light wave is refracted.
7. it absorbs light of other wavelengths but does not absorb red light; the red light is transmitted, so the solution is transparent (at least at some wavelengths)
8. it absorbs light from the red and violet ends of the spectrum, and reflects the light that is not absorbed (green)

Answers to Worksheet 3.6.7

1. waves; wavelengths; human eye; spectrum
2. a) A = 3.9 cm; B = 3.0 cm; C = 2.7 cm

   b) If students calculate the ratios of 'typical' (mid-range) red, green and blue light waves, they will see that the ratios are very similar to those calculated from the drawings.
3. a) Diagram should show red and blue rays absorbed, and only the green ray transmitted.

   b) Diagram should show green and blue rays absorbed, and the red ray reflected.

Answers to Practical sheet 3.6.7

6. a) A continuous spectrum is seen, with the red end opposite the edge of the mirror resting on the bottom of the trough.

   b) the angle should be about 30°
7. Results: the red sheet lets only light in the red region of the spectrum through; green, the green region; blue, the blue region; similar results for other coloured plastic sheets.

6.8 Applying key ideas

Rainbows

Objectives

- Extract ideas from the text about light and its properties.
- Use information about reflection and refraction to explain the formation of rainbows.
- Apply the ideas to explain the way CDs and DVDs can produce coloured light.

Outcome

- Clear and effective responses to questions, indicating understanding and the next steps in learning.

The purpose of this activity is to provide an opportunity to see how successfully students are grasping the key ideas so far. It is not designed to be used as a formal test – it might be that the students work on the questions collectively. It does provide an opportunity for you to look at written work, engage students in discussions and form ideas about progress being made.

The tasks are progressive. Lower-attaining students should be able to tackle the first two or three tasks and higher-attaining students will find the later tasks more challenging.

Teaching and learning

Engage

- Challenge the students to **reflect back** over the last few lessons and **identify** some of the key ideas they have learned about and investigated. Draw out some responses about these ideas, and explain that it is important that, as learners, they develop a sense of their own progress.
- You may wish to repeat (it may have been shown in previous lesson) a demonstration of the formation of a 'rainbow' from a spray of water on a sunny day.
- Ask the students, working individually, to read the text about rainbows in the Student Book. Point out the photograph of the double rainbow (Figure 3.6.8b) and encourage the students to look for similarities and differences between the inner and outer bows.

Challenge and develop

- Ask the students, working individually, to complete the first two tasks in the Student Book. Depending on their ability to interpret scientific diagrams, you may need to adopt strategies such as:
  - asking the students to **redraw** the diagram, **explaining** what each line they draw represents
  - working with a group of students who have difficulty verbalising what the diagram shows
  - allowing the students to work in pairs to **share** ideas and **interpret** the diagram.

Explain

Ask the students to attempt tasks 3 and 4 in the Student Book, either individually or collaboratively. Again, depending on their ability to work with and interpret scientific diagrams you may need to adopt the strategies listed above.

Consolidate and apply

- Ask the students to **present** their responses to the various tasks, either orally or by **displaying** their work.

Extend

- Ask students able to progress further to attempt task 5 in the Student Book. It is a challenging task, and even those who have progressed quickly and correctly may need some guidance.

Plenary suggestions

How well did we do? Ask the students for responses to questions such as:

- How well were you able to approach those tasks?
- Did you manage to present some high-quality responses?
- Are there some ideas that you need to strengthen in the later part of this topic?

If there are, make a note.

Provide the class with some overall feedback, indicating ideas that they have grasped effectively and those that may need developing further.

Likely responses and next steps in learning

1. *Making a rainbow* gives the students an opportunity to assimilate information from the text and use it to write their own descriptions of the conditions needed for a rainbow to be formed.
2. *Colours of the rainbow* gives the students an opportunity to look at a diagram and to describe the spectrum that was formed, noting that red is on the outside and that blue is on the inside (nearer to the ground). They are then required to show their understanding of the ray diagram by explaining the formation of the observed spectrum. This includes demonstrating that they can use the words 'refraction' and 'reflection' correctly, though they are not, at the moment, required to explain why refraction and reflection happen.

   At points 1 and 3 refraction happens; at point 2 reflection happens. The colour of light wave A is violet and that of light wave B is red (notice how they cross over in the water droplet).
3. *Explaining direction changes* requires the students to recall what they have learned about the properties of light and to apply this to explain that refraction happens at the interface between water and air – and that this causes the change in direction of the light waves. The students will also need to apply their knowledge of reflection to explain the 'bouncing back' of light waves within the water droplet.

   The explanation needs to include the idea that light waves have a range of wavelengths and frequencies, and that the amount of diffraction depends on wavelength.
4. *Double rainbows* requires the students to look at a photograph (Figure 3.6.8b in the Student Book) and to describe the different spectra of the inner bow (red is on the outside) and the outer bow (blue is on the outside). They need to demonstrate their deeper understanding of refraction and reflection by applying their knowledge to explain the formation of a double rainbow and the reason for the difference in the two spectra.

   This is similar to task 3, but there is a double internal reflection in the outer bow, which results in the upper wave being red and the lower one being violet – the oppostite way around to the inner bow.
5. *CDs, DVDs and rainbows* requires the students to consider ideas used throughout the topic so far. Their understanding is checked as they explain why CDs and DVDs can be used to produce a spectrum.

6.9 Understanding energy transfer and change

Lesson overview

Learning objectives

- Describe the ways in which energy is stored.
- Describe the ways that energy can be transferred from one store to another.
- Explain that any change – physical or chemical – results in a transfer of energy.

Learning outcomes

- Recall ways in which energy is stored. [O1]
- Recall ways in which energy is transferred from one store to another. [O2]
- Explain that energy transfer occurs when physical and chemical changes happen. [O3]

Skills development

- Thinking scientifically: ask questions
- Working scientifically: develop explanations
- Learner development: develop resilience

Resources items such as a ball, ice, water, spring, elastic band, matches, and paper may be required, depending on the starter activity; Worksheet 3.6.9

Digital resources Quick starter; Interactive activity: Match the description to the type of energy store; Video

Common misconceptions The transfer of energy *causes* the change.

Key vocabulary energy store, energy transfer

Teaching and learning

Engage

- Demonstrate a range of physical and chemical changes, such as holding a ball in the air and dropping it, melting ice (or other suitable solid), boiling water, stretching and releasing a spring or elastic band, striking a match, and burning a piece of paper. Ask the students to **identify** what changes are happening and whether or not there is evidence of energy being transferred during the change. [O1]

Challenge and develop

- **Pairs/small groups** After the demonstrations, introduce the idea of storing energy. Ask the students to read the 'Storing energy' section in the Student Book and to answer questions 1–3. [O1]
- **Individual work** Ask the students to complete task 1 of Worksheet 3.6.9. [O1]
- Raise the question of how energy might be transferred when one energy store changes into another store – like when ice becomes liquid or a stretched elastic band is no longer stretched. [O2]

Explain

- Discuss with the students the evidence for energy being transferred during change. [O1&2]
- Ask the students to read the 'Types of energy transfer' section in the Student Book and to answer questions 4–6. [O2]
- Review the changes demonstrated at the beginning of the lesson, and ask the students to **name** the energy stores and to say, in each case, how energy was transferred during the change. [O1&2]

Consolidate and apply

- Check the students' understanding of energy stores and energy transfers by asking them to complete task 2 of Worksheet 3.6.9. [O1&2]

- Ask the students to read the 'Explaining change and energy transfer' section in the Student Book. When they are sure that they **understand** the diagram in Figure 3.6.9c, ask them to complete task 3 of Worksheet 3.6.9. [O3]
- Ask the students to answer questions 7 and 8 in the Student Book. [O3]

Extend

- Ask students who are able to progress further to **think about** changes that result in energy being stored in two or more different ways. Ask them how the relative proportions of energy in each store might be represented diagrammatically. The idea is to prompt the students to **consider** conservation of energy and the use of Sankey diagrams. [O1, 2&3]

Plenary suggestions

Describing energy transfers Organise the students into small groups. Ask each group to **describe** three changes – they must not use any of the changes that they have already met in the lesson – naming the stores before and after, and the way in which energy was transferred. Ask each group to read out its changes, **describing** what the stores and types of energy transfers are. After all the groups have presented their answers, ask the class to vote on which group described the changes correctly and most clearly. [O1, 2&3]

Answers to Student Book questions

1. thermal; chemical; gravitational
2. any one from: magnetic; kinetic; elastic
3. thermal; chemical; gravitational; kinetic; elastic; magnetic
4. electrically; mechanically; through heating; by radiation
5. conduction
6. by air particles moving as (longitudinal) waves
7. by radiation (light waves transfer energy); and by heating the surroundings
8. physical and chemical changes produce energy transfers; energy transfers do not cause change

Answers to Worksheet 3.6.9

1. *Across*: 1 chemical 4 thermal 6 kinetic
   Down: 2 magnetic 3 gravitational 5 elastic
2. a) kinetically
   b) an electric current
   c) energy; thermally
   d) transferred; electromagnetic waves
3. a) *Initial energy stores*: red hot iron; cold water; barrel. *Final energy stores*: cold iron, hot water, hot barrel. *Energy is transferred*: thermally (by conduction).
   b) *Initial energy stores*: cold water; cold beaker; natural gas; oxygen. *Final energy stores*: boiling water; hot beaker; products of combustion (carbon dioxide and water). *Energy is transferred*: thermally (by heating).
   c) *Initial energy stores*: the Sun; stationary car (with solar panel). *Final energy stores*: moving car; *Energy is transferred*: by radiation to the solar panel; and electrically to move the car.

6.10 Explaining thermal conduction and radiation

Lesson overview

Learning objectives

- Describe the warming and cooling of objects.
- Explain the relationship between energy transfer and temperature change.
- Compare the transfer of energy by thermal conduction and by radiation.

Learning outcomes

- Recall that warming causes a rise in temperature and that cooling causes a drop in temperature. [O1]
- Explain that energy is transferred from a warm object to a cooler object until both are at the same temperature. [O2]
- Explain that conduction requires a medium to transfer energy, but that radiation can transfer energy through empty space. [O3]

Skills development

- Thinking scientifically: analyse data
- Working scientifically: interpret evidence
- Learner development: collaborate effectively

Resources graph paper; equipment and materials as detailed in the Technician's notes; Worksheet 3.6.10; Practical sheet 3.6.10; Technician's notes 3.6.10

Digital resources Quick starter; Slideshow: Energy transfers around the home; Interactive activity: Drag the statements and descriptions into the correct group - energy transfer by conduction, or energy transfer by radiation

Common misconceptions Putting on a coat on makes you warmer – the coat transfers energy to your body.

Key vocabulary temperature, thermal equilibrium, thermal conductor, thermal insulator, infrared radiation

Teaching and learning

Engage

- Ask the students to **describe** some situations in which they feel cold, and what they do to keep warm. Even if they do not use the correct terminology at this stage, ask them if they can group ways of keeping warm. Answers are likely to fall into three categories – standing close to a fire or electric heater (radiation); putting their hands on something warm (conduction); or putting something on that slows down energy transfer from the body to cold surroundings (insulation). [O1&2]

Challenge and develop

- Ask the students to read the 'Cooling and warming' section in the Student Book. Take some time to look at Figure 3.6.10b with the students. If possible, display the image to the whole class and point out different features of the graph. If time allows, the experiment could be demonstrated, with students being invited to **take temperature readings**.
- Next, ask the students to answer questions 1 and 2 in the Student Book. At this stage, explanations are not required. [O1&2]
- **Group work** Ask the students to complete task 1 of Worksheet 3.6.10. If necessary, remind the students how to set about solving a 'fallen phrase' puzzle. They could work in teams of four, with each pair in the team tackling one of the puzzles. [O1&2]

Explain

- Ask the students to answer questions 3–5 in the Student Book and to complete task 2 of Worksheet 3.6.10. [O2]

- **Group work** The students can then **carry out** the practical work, following the instructions on Practical sheet 3.6.10. Ideally they need to work in groups of six. Each group has a metal beaker, a plastic beaker and an expanded polystyrene cup – the group divides into three pairs and each pair is responsible for **investigating** one of the containers and for **drawing a graph** of temperature against time. The three pairs of students in a group can then **compare** the results from the three different containers. If available, a datalogger could be used and the graph generated from an Excel (or similar) spreadsheet. It depends on whether or not the activity is intended to provide an opportunity to practice drawing graphs. [O1&2]
- Ask the students to read the 'Conduction and radiation' section in the Student Book and to answer questions 6–8. [O3]

Consolidate and apply

- Revisit Figure 3.6.10b in the Student Book. Remind the students about the particle model and changes of state. Ask them if they can **explain** the shape of the cooling curve. [O1&2]
- Using prompts if necessary, draw out the idea that the two downward curves show that energy is being transferred from the water (and then ice) to the ice/sodium chloride mixture. The plateau indicates that solidification is occurring, which releases energy as bonds form between particles. [O1&2]

Ask higher-attaining students to complete task 3 of Worksheet 3.6.10. [O3]

Extend

- Ask students able to progress further to read the 'Did you know?' box in the Student Book and to **explain** how double glazing helps to keep rooms warm in the winter *and* cool in the summer.

Plenary suggestions

Highlighting uncertainties Ask the students to say how well they thought they coped with the tasks on Worksheet 3.6.10, and to write on sticky notes anything that they are still unsure about. [O1, 2&3]

Fallen phrases If time was short earlier on, the fallen phrase puzzles could be completed (task 1 of Worksheet 3.6.10). [O1, 2&3]

Answers to Student Book questions

1. the higher its temperature, the hotter something is
2. The water cools until the temepratur reaches 0 °C. Ice begins to form and, while this is happening, the temperature remains at 0 °C. When the water has completely solidified, the temperature of the ice begins to fall below 0 °C.
3. The temperature of the water rises as energy is transferred from the hot can to the water; the temperature of the can decreases. Eventually both can and water will be at the same temperature.
4. Energy is transferred from the metal to the water.
5. An 'open system' is one in which all the surroundings (kitchen, everything in it and everything outside of it) are taken into account.
6. A conductor transfers energy through itself quickly; an insulator transfers energy only very slowly.
7. Conduction needs large numbers of particles to be touching, so that energy can be transferred between them. Gas particles are spread out; they do collide, but they are not in constant contact with one another.
8. They vibrate more rapidly about their positions in the structure.

Answers to Worksheet 3.6.10

1. a) graph showing how temperature decreases with time

   b) measuring how quickly temperature increases or decreases
2. a) something to hold the rod (e.g. a clamp) and something with which to heat the rod (e.g. a Bunsen burner)

   b) The rod will get very hot, and should be clamped and not touched by hand.

   c) Place thermistors at regular intervals along the rod; heat the rod at one end using a Bunsen flame; record the temperature measured by each thermistor over a period of time.

   d) For each thermistor, plot temperature against time on the same axes, labelling each series of points to show how far the thermistor was from the heated end of the metal rod.
3. Shade is produced when materials absorb light waves, so energy carried by light waves (radiation) cannot directly reach an object in shade. However, radiated energy is transferred between air particles by conduction and convection. So it is still possible to feel warmth from the Sun in the shade on a sunny day, but it will not feel as warm as it does in direct sunlight.

6.11 Understanding energy transfer by fuels and food

Lesson overview

Learning objectives

- Describe the use of fuels in the home.
- Explain that foods are energy stores and that the amount stored can be measured.
- Explain that energy is not a material and can be neither created nor destroyed.

Learning outcomes

- Recall the types of fuel used in the home. [O1]
- Explain that foods store different quantities of energy that can be measured. [O2]
- Explain that when energy is transferred from fuels and food, the total amount of energy before and after remains the same; it is just stored differently. [O3]

Skills development

- Thinking scientifically: use units and nomenclature
- Working scientifically: evaluate data
- Learner development: ask questions

Resources food labels; various electrical appliances with their power ratings showing; Worksheet 3.6.11 (the final page copied onto card)

Digital resources Quick starter; Interactive activity: Drag the units into the correct group - unit of energy, or not a unit of energy; Video

Key vocabulary fuel, kilowatt-hour (kWh), calorie, metabolism, Law of Conservation of Energy

Teaching and learning

Engage

- Ask the students how their home is heated, how they get hot water at home and which electrical appliances are in their home. This is not a 'competition' and needs to be approached sensitively. An alternative might be for the students to ask the question of other teachers in school, but the students will need advance warning, e.g. during the previous lesson. It might also help if teachers are primed that this will happen. [O1]

Challenge and develop

- Ask the students to read the 'Fuels and energy' section in the Student Book and to answer questions 1–3. [O1]
- **Pair work** When they have answered the first three Student Book questions, ask the students to complete task 1 of Worksheet 3.6.11. This might be done competitively with students working in pairs to **solve** the puzzle. When they have finished they can hand in their answers and when all have finished, the first correct answer to have been handed in is identified. [O1]
- Review with the students what they learned about the energy content of food in their biology lessons. If possible it would be helpful to have a variety of food packages and labels to look at. Discuss how energy is stored in the body. [O2]
- Ask them to **recall** earlier work and to state how the energy content of food is measured. [O2]

Explain

- Ask the students to answer questions 4–6 in the Student Book and to complete task 2 of Worksheet 3.6.11 when they have finished. The final question in the task could be part of a whole class discussion about energy intake and expenditure. Make sure that the students **understand** the central idea of transfer, together with the concept of conservation of energy. [O2]

- Remind the students of previous lessons in which they learned about energy stores and transfers. Ask them to **give some reasons** why the rate of energy transfer is important. [O2]
- Ask the students to read the 'Conservation of energy' section in the Student Book and to answer questions 7 and 8. [O3]

Consolidate and apply

- Ask the students to complete task 3 of Worksheet 3.6.11. [O3]

Extend

- Ask students able to progress further to **think of** different ways that transferred energy is stored. Ask them to look again at the Sankey diagrams on Worksheet 3.6.11 and to **consider** what is meant by the terms 'useful energy' and 'wasted energy'. This will re-introduce the idea that not all of the energy is transferred in a desired change – for example an electric light also generates heat. The students might be asked to **discuss** what they think is meant by the 'efficiency' of an electrical appliance. [O3]

Plenary suggestions

Summaries Arrange the students in pairs. Ask each pair to **write a sentence** that summarises something that they have learned in the lesson. Encourage pairs to **share** their statement with the class. [O1, 2&3]

Answers to Student Book questions

1. gas; coal; oil; wood
2. electricity meter
3. a) kilowatt-hours (kW h) b) cubic metres (m^3)
4. The body makes compounds from food, which act as energy stores in the body's cells. Energy is transferred in the process of breaking down these compounds.
5. it helps people to choose healthy diets
6. total energy content of a serving = 310 calories = $\frac{310}{240} \times 1000 = 1.3$ kJ

   the serving size is 144 g, so the energy in 100 g = $\frac{1.3}{144} \times 100 = 0.9$ kJ
7. any suitably described physical or chemical change; for example heating water in a kettle, where energy is transferred to the water, the kettle and the surroundings; or burning a candle, where energy is transferred to the products of combustion (carbon dioxide and water) and to the surroundings
8. any correct sentence(s) that convey(s) the idea that the total energy stored before a change equals the total energy stored after the change

Answers to Worksheet 3.6.11

1. The use of gas and electricity in homes is measured using meters.
2. a) *Very low*: 0–2.52 kJ/g; *Low*: 2.52–6.30 kJ/g; *Medium*: 6.30–16.80 kJ/g; *High*: 16.80–37.80 kJ/g

   b) i) 2360.4 kJ ii) 3927.0 kJ

   c) if energy intake exceeds energy expenditure, the body's food store increases, and so does body mass
3. a) total energy transferred = 90 kJ (heating surroundings) + 10 kJ (light) = 100 kJ = energy transferred electrically

   b) i) widths: fuel = 4.3 cm; environment = 2.5 cm; home = 1.4 cm; machinery = 0.2 cm; during transmission = 0.2 cm

   ii) total widths for energy tranferred = 4.3 cm (= width of energy stored in fuel)

   iii) Because the widths are proportional to the energy (transferred in different ways), the total energy transferred = the energy stored in the fuel; so the Law of Conservation of Energy holds true.

6.12 Comparing rates of energy transfer

Lesson overview

Learning objectives

- Describe what is meant by 'rate of energy transfer'.
- Recall and use the correct units for rate of energy transfer.
- Calculate quantities of energy transferred when change happens.

Learning outcomes

- Recall that the rate of energy transferred is calculated as the quantity transferred divided by the time taken for it to be transferred. [O1]
- Use the watt as the unit of energy transfer in calculations. [O2]
- Calculate quantities of energy transferred using power ratings and time measurements. [O3]

Skills development

- Thinking scientifically: analyse data
- Working scientifically: evaluate data
- Learner development: communicate effectively

Resources sticky notes; 100 cm^3 and 400 cm^3 beakers; cotton wool (or other insulating material); thermometer; hot water; a number of electrical/electronic devices that have their power ratings shown; Worksheet 3.6.12

Digital resources Quick starter; Interactive activity: Match the calculated quantity to the correct description

Key vocabulary power, watt, kilowatt

Teaching and learning

Engage

- Ask the students to **write** on a sticky note one reason why the speed that a quantity of energy is transferred at might need to be controlled, **giving** an example to illustrate. Ask them, one by one, to stick their notes on a wall. As each new note is added, rearrange them into groups containing similar reasons and examples. [O1]

Challenge and develop

- It is likely that one group of sticky notes will be about keeping things warm by slowing down the rate of energy transfer. Demonstrate this by showing the difference in how quickly water cools when it is in a glass beaker compared with how long it takes to cool when it is in a glass beaker that is placed in a larger beaker with insulating material packed in the gap between the two. [O1]
- Ask the students to read the 'Rate of energy transfers' section in the Student Book and to answer questions 1–3. Some may find the everyday use of the word 'power' and its scientific meaning a little confusing. [O1]
- Ask the students to complete task 1 of Worksheet 3.6.12. [O1]
- Raise the question of what units might be used to measure the rate of energy transfer. [O2]

Explain

- Following on from the idea of units, introduce the watt as the quantity of energy (joules) transferred per second. Arrange a variety of electrical/electronic devices so that the students can move around the room to look at them. Ask the students to **write down** what they are and their power ratings. [O2]
- Reinforce the use of units by asking the students to answer questions 4–6 in the Student Book. [O2]
- Ask them to complete task 2 of Worksheet 3.6.12. The students could work in pairs, especially for part c). [O2]
- Follow the work on comparing rates of energy transfer by asking the students to read the 'Quantities of energy transferred' section in the Student Book and to answer questions 7 and 8. [O3]

Consolidate and apply

- **Pairs** Ask the students to complete task 3 of Worksheet 3.6.12. [O3]
- Encourage the students to **share** with the class the benefits and disadvantages they listed for using standby modes on electronic devices and gadgets. [O3]

Extend

- Ask students able to progress further to look at the information from the Energy Saving Trust's website and from its report 'Powering the nation – household electricity-using habits revealed', which are detailed on Worksheet 3.6.12. Ask them to **explain** why turning off devices, rather than just leaving them on standby, might reduce the energy consumed by up to 16%. [O3]

Plenary suggestions

Learning triangle Ask the students to draw a learning triangle (a large triangle with a smaller inverted triangle that just fits inside it, so there are four triangles in total). Ask them to think about the lesson and to write in the outer three triangles something that they've seen, something that they've done and something that they've discussed. Then ask them to add in the central triangle something that they've learned. [O1, 2&3]

Answers to Student Book questions

1. the quantity of energy transferred per second, minute or hour
2. the faster the rate of change, the faster the rate of energy transfer
3. keeping warm in cold weather means slowing the rate at which energy is transferred from the body to the cold air; wearing layers of clothing slows the rate of energy transfer (the materials and layers of air between them act as insulators)
4. 20 J/s
5. microwave oven; toaster; electric oven
6. the calculation gives 107.5; there is no such thing as half a laptop so the answer is 107
7. 216 kJ
8. 4500 J (4.5 kJ)
9. a) 2 592 000 J (2592 kJ) b) 864 000 J (864 kJ)

Answers to Worksheet 3.6.12

1. a) decreases; stored; conduction

   b) all curves should start at 80 °C and show that the rate of cooling slows down until they reach the same room temperature

   i) The graph for the *copper can* should have the steepest initial curve, showing the fastest rate of cooling.

   ii) The graph for the *glass beaker* should be the next steepest.

   iii) The graph for the *polystyrene cup* should be the least steep.

   The difference in the steepness of the initial curve should be bigger between copper and glass, than it is between glass and polystyrene.
2. a) the rate of transfer of energy b) 65 J

   c) i) E has a higher power rating than A.

   ii) The power can be changed from the lowest value (defrosting) to the highest value (high).

   iii) measure the temperature rise in a given volume of water heated for a specified length of time; calculate the energy transferred in a certain time for each appliance, a number of times
3. a) energy needed = 4.5 × 300 × (100 – 20) = 100 800 J (or 100.8 kJ)

   b) time taken = 108 000 ÷ 2000 = 50.4 seconds; it will take a little longer because some energy is transferred to the kettle itself and to the surroundings; the faster the kettle boils, the lower this unwanted energy transfer will be

   c) The summary needs to highlight the convenience of the standby mode, e.g. easy to pick up the remote control and turn the TV on; but constrast this with the extra energy consumed and paid for while devices are in standby mode.

6.13 Looking at the cost of energy use in the home

Lesson overview

Learning objectives

- Describe the information a typical fuel bill provides.
- Explain and use the units used on a fuel bill.
- Explain how the cost of energy used can be calculated.

Learning outcomes

- Recall the type of information given on a fuel bill. [O1]
- Explain units of energy and how they are converted from one to another. [O2]
- Calculate the cost of energy from information about power and time. [O3]

Skills development

- Thinking scientifically: evaluate risks
- Working scientifically: design investigations
- Learner development: communicate effectively

Resources Worksheet 3.6.13

Digital resources Quick starter; Interactive activity: Match the descriptions to the numbers and units found on household electricity bills; Hangman: Key vocabulary game

Common misconceptions Energy is consumed in homes, which implies that it is being used up.

Key vocabulary unit of electricity, kilowatt-hour (kWh)

Teaching and learning

Engage

- Ask the students to **think about** previous lessons and to **write down** the units used for the quantity of energy and the rate of energy transfer (power). [O1&2]

Challenge and develop

- Ask the students to look at Figure 3.6.13a in the Student Book. Read the 'Fuel bills' section of the Student Book as a class and then ask the students to work in pairs to answer questions 1–4. [O1&2]
- **Pair work** Next, ask the students to work in their pairs and read through the two sections 'Power of domestic appliances' and 'Calculating the cost of energy used' in the Student Book. They should then answer Student Book questions 5–9. [O2&3]

Explain

- To reinforce their answers to Student Book questions 1–4, ask the students to complete task 1 of Worksheet 3.6.13. [O1&2]
- Similarly, to reinforce their learning and their answers to Student Book questions 5– 9, ask the students to complete task 2 of Worksheet 3.6.13. [O2&3]

Consolidate and apply

- **Pair work** The lesson has focused on energy calculations. To draw many of the ideas together, ask the students to work in pairs to complete task 3 of Worksheet 3.6.13. [O2&3]

Extend

- Ask students able to progress further to **think about** the relative costs of electricity and gas, and to **explain** why electricity is more expensive than gas. [O3]

Plenary suggestions

Highlighting uncertainties Ask the students to **write down** one thing about energy (quantities, rates of transfer and use in the home) that they are sure of, one thing that they are unsure of and one thing that they need to know more about – they should be specific. Then, working in groups of five or six, ask the students to **share** and **discuss** what they have written. This discussion may lead to some students gaining a better understanding of previous uncertainties, and it may also lead to new uncertainties emerging. After the discussion ask each group to **share** what they is confident about, what they are less sure about and what they need to know more about. Draw these responses together to identify revision topics. [O1, 2&3]

Answers to Student Book questions

1. using electricity meters; and gas meters
2. one joule is a very small amount of energy; the numbers would be too large to write on a bill
3. a) £17.36 b) £19.24
4. A standing charge pays for fixed costs, e.g. keeping the home connected to the network, meter reading, maintenance, emergency gas supplies, and Government initiatives such as carbon-reduction commitments and supporting vulnerable households. Students do not need to know all of these, just that the energy costs to a consumer are more than the fuel itself.
5. hair dryer
6. *Energy* is a quantity stored in different ways. *Power* is a measure of how quickly energy is transferred during a change.
7. a) 336 000 J

   b) 336 kW ÷ (60× 60) h = 0.093 kW-h; cost = 0.093 × 13p = 1.2p
8. £1428

Answers to Worksheet 3.6.13

1. a) meter 1 is for electricity (units kWh); meter 2 is for gas (units m^3)

   b) meter 1: 6543 kWh; meter 2: 8634 m^3

   c) an energy company has fixed costs that are the same no matter how much energy is used, e.g. maintenance work, billing costs; these are covered by the standing charge; the other cost is for the fuel consumed, which varies from home to home and from month to month
2. a) 4.2 J

   b) i) 10 080 000 J ii) 10 080 kJ

   c) 36.4 p d) 11.2 p
3. Students are not expected to give a detailed method. Equal volumes of water must be used, and the temperature rise must be the same in each case. Suitable suggestions should be made for measuring the quantity of energy transferred by electric current and by burning gas (the volume of gas used would need to be measured). There should be a considered safety assessment. Repeat data should be obtained to check reliability.

6.14 Checking students' progress

The 'Checking your progress' section in the Student Book indicates the key ideas developed in this chapter and shows how students progress to more complex levels. It is provided to support students in:

- identifying those ideas
- developing a sense of their current level of understanding
- developing a sense of what the next steps in their learning are

It is designed either to be used at the end of a chapter to support an overall view of progress, or alternatively during the teaching of the chapter. Students can self assess or peer assess using this as a basis.

It would be helpful if students could be encouraged to provide evidence from their understanding or their notes to support their judgements. In some cases it may be useful to explore differences in the descriptors for a particular idea so that students can see what makes for a 'higher outcome'.

It may be useful with some descriptors to provide examples from the specific work done, such as an experiment undertaken or an explanation developed and recorded. If marking and feedback uses similar ideas and phrases, this will enable students to relate specific marking to a more general sense of progress.

To make good progress in understanding, science students need to focus on these ideas and skills:

Students who are making modest progress will be able to:	Students who are making good progress will be able to:	Students who are making excellent progress will be able to:
Recognise that energy may be transferred by different types of waves, and know the difference between longitudinal and transverse waves.	Explain wave oscillation, the reflection and superposition of waves, and the terms frequency and wavelength.	Compare the properties of sound waves, waves in water and light waves.
Recognise that light can be reflected by some materials and absorbed by others.	Explain how some materials absorb energy, and the differences between transparent, translucent and opaque materials.	Use diagrams to explain the difference between diffuse and specular reflection.
Describe the ray model of light using the idea that light travels in straight lines.	Explain the difference between reflection and refraction, and describe what happens when light waves are refracted.	Use ray diagrams to explain how a pinhole camera and the eye work.
Recognise that various effects can occur when materials absorb light, for example chemical reactions or a flow of electric current.	Explain how the transfer of energy carried by light happens during photosynthesis in plants and by electron release in photovoltaic cells.	Explain the formation of photochemical smog.
Describe the formation of a spectrum from white light.	Explain how white light can be split into a continuous spectrum of colours, called the visible spectrum.	Use the concepts of reflection and absorption of light to explain why some materials (transparent, translucent and opaque) are coloured.
Describe different ways in which energy can be stored and different ways in which energy can be transferred.	Explain that energy is transferred from one type of energy store to another when change happens, and understand that energy transfer does not cause change.	Explain that all changes, physical or chemical, result in a transfer of energy.
Describe the transfer of energy by heating and cooling.	Explain the relationship between energy transfer and temperature difference.	Compare the transfer of energy by conduction and by radiation.
Recall the units used to measure quantities of energy, including joules, calories and kilowatt-hours.	Explain that energy can be neither created nor destroyed (the Law of Conservation of Energy).	Carry out calculations of quantities of stored and transferred energy.
Describe what is meant by rate of energy transfer.	Identify the rate at which electrical appliances transfer energy (their power rating), using the correct units (watts or kilowatts).	Compare rates of energy transferred when electrical appliances are used.
Explain the data given on an energy bill, including the units used for energy 'consumed' (transferred to appliances in the home) and the meaning of 'standing charge'.	Use the power rating of an appliance to calculate the amount of energy transferred.	Calculate the cost of energy used in different scenarios.

6.15 Answers to Student Book Questions

This table provides answers to the Questions section at the end of Chapter 6 of the Student Book. It also shows how different questions assess attainment in terms of the focus and style of a question as well as the context. Question-level analysis can indicate students' proficiency in approaching different aspects of scientific understanding and different types of answer.

Q	Answer		Focus			Style			Context	
		Marks available	Knowledge & understanding	Application	Evaluation of evidence	Objective test question	Short written answer	Longer written answer	Light waves	Energy stores and transfer
1	b	1	x			x			x	
2	a	1	x			x			x	
3	b	1	x			x			x	
4	c	1	x			x			x	
5	• chemical change • flow of electric current	1 1	x x				x x			x x
6	reasonable example, such as: • energy stores; battery and lamp • energy transferred; electrically	 1 1	 x x				 x x			 x x
7	any **four** of: • price of a unit of energy • one unit = 1 kilowatt-hour • electricity metered in kilowatt-hours • gas metered in cubic metres • number of units consumed • cost of energy used • standing charge	4	x					x		x
8	suitably completed diagram that traces the refracted ray back to where the eye perceives the object's origin to be	1		x			x		x	
9	the different-colored pieces of glass are absorbing and reflecting light from different parts of the visible spectrum	1		x			x		x	
10	an electric kettle needs to transfer energy more rapidly	1		x			x			x
11	16 800 J (or 16.8 kJ)	1		x			x			x

Q	Answer		Focus			Style			Context	
		Marks available	Knowledge & understanding	Application	Evaluation of evidence	Objective test question	Short written answer	Longer written answer	Light waves	Energy stores and transfer
12	• as the archer pulls back and stretches the string, energy is transferred from stores in the body to the stretched string	1		x				x		x
	• when released, the arrow shoots forward. Energy stored elastically in the string is transferred to the moving arrow, where it is stored kinetically	1		x				x		x
13	a) the rate of transfer slows down	1		x			x			x
	b) the rate of energy transfer would be slower; the curve would have a similar shape but be less steep	1		x			x			x
14	• dirt and dust form an uneven layer on the surface of the car; some light waves are absorbed, others are reflected diffusely because of the irregular surface	2		x				x	x	
	• removing the dirt and dust, followed by polishing, provides a flat surface; some light waves are absorbed, but those that are reflected bounce back in the same direction as one another (specular reflection)	2		x				x	x	
15	chlorophyll in the leaf absorbs light from the red and blue regions of the spectrum; and reflects light from the green region	1			x			x	x	
16	The uneven surface becomes smooth where the sticky tape is, and so all light waves refract in the same direction when they pass from air, through the tape and into the glass. This means that you can see that the frosted glass with tape on it is transparent	1			x			x	x	
17	• for equal quantities of the types of milk, the total energy stored decreases from whole (full cream) to skimmed milk	1			x			x		x
	• milk contains three types of substances: protein, carbohydrate and fat	1			x			x		x
	• the masses of protein and carbohydrate per 100 ml are very similar for all three types of milk	1			x			x		x
	• this suggests that differences in energy stored are due to the quantities of fat in milk	1			x			x		x
	Total possible	**30**	**12**	**12**	**6**	**4**	**10**	**16**	**12**	**18**

Scheme of Work

Teaching Key Stage 3 in three years

The following pages show the full plan of the scheme of work for Collins Key Stage 3 Science Teacher Pack 3 taught over 3 years.

Teaching Key Stage 3 in two years

If you are using the Collins KS3 Science scheme to deliver the Programme of Study in two years there are three ways you can do it:

1. Focus on the lessons shaded in the table. By so doing you will have visited all the key ideas.
2. Use the shaded lessons as a starting point but draw on ideas, activities and questions as necessary, i.e., 'swapping out' the occasional activity on an indicated lesson.
3. Use the introductory lesson and/or the 'Applying key ideas' lesson to see what students are more confident with and what time would be better spent on.

Collins Connect is our digital learning platform that offers a range of linked resources to enhance your lessons.

Chapter 1: Variation for Survival

Lesson	Lesson title	Overarching objectives	Learning objectives	CD-ROM resources	Collins Connect resources	Notes for two-year scheme
3.1.2	Exploring differences	Differences between species The importance of maintaining biodiversity	• Identify differences between different species • Explain the importance of diversity	Worksheet 3.1.2a; Worksheet 3.1.2b	Quick starter; Slideshow: The five kingdoms – protoctists, prokaryotes, fungi, plants and animals; Interactive activity: Drag the animals to the correct phyla; Slideshow: Life in different environments – a look at how living things adapt to extreme environments; Slideshow: Hybrids: Definition and examples; Video	
3.1.3	Looking more closely at variation	The variation between individuals within a species being continuous or discontinuous, to include measurement and graphical representation of variation	• Explain the difference between continuous and discontinuous variation • Investigate variation within a species • Evaluate the importance of variation in organisms	Worksheet 3.1.3	Quick starter; Interactive activity: Drag the statement to the correct correlation	Refer to differences between species and the importance of maintaining biodiversity, from 3.1.2

3.1.4	Exploring the causes of variation	Heredity as the process by which genetic information is transmitted from one generation to the next	• Identify some features of organisms that are inherited and some that are determined by their environment • Understand that offspring from the same parents may show considerable variation • Evaluate the importance of genetic and environmental variation to the survival of the organism	Worksheet 3.1.4; Practical sheet 3.1.4; Technician's notes 3.1.4	Quick starter; Slideshow: The causes of variation: A look at genetic and environmental factors; Interactive activity: Drag the characteristics to the correct group - caused by genetic factors, environmental factors, or both; Video	
3.1.5	Learning about selective breeding	Heredity as the process by which genetic information is transmitted from one generation to the next	• Describe how selective breeding can produce organisms with desirable characteristics • Explain the process of selective breeding • Evaluate the importance of selective breeding, and explore the ethical issues involved	Worksheet 3.1.5	Quick starter; Interactive activity: Reorder the sentences to describe the process of selective breeding of cattle	
3.1.6	Finding out how organisms survive	The variation between species and between individuals of the same species means some organisms compete more successfully, which can drive natural selection	• Describe how variation causes competition for resources, and drives natural selection • Explain the theories of Lamarck, Wallace and Darwin • Evaluate the importance of Darwin's work	Worksheet 3.1.6; Practical sheet 3.1.6	Quick starter; Interactive activity: Reorder the sentences to describe the process of evolution by natural selection; Slideshow: How life on Earth evolved: A look at Charles Darwin's theory of evolution; Hangman: Key vocabulary game	

3.1.7	Applying key ideas		• Extract ideas about variation within a species from the text, including earlier topics • Apply ideas about variation to explain evidence • Apply ideas and information about selective breeding to propose the outcome of a process	Worksheet 3.1.7		
3.1.8	Understanding why siblings are different	Heredity as the process by which genetic information is transmitted from one generation to the next	• Identify inherited features in plants and animals that vary between offspring • Explain how inherited differences arise by genetic material from both parents combining • Describe how identical twins occur and analyse data about their features	Worksheet 3.1.8; Practical sheet 3.1.8	Quick starter; Video; Interactive activity: Drag the statements about twins to the correct group	
3.1.9	Looking inside a cell's nucleus	A simple model of chromosomes, genes and DNA in heredity, including the part played by Watson, Crick, Wilkins and Franklin in the development of the DNA model	• Identify that the nucleus contains chromosomes which carry inherited genetic information • Explain that chromosomes are made of genes containing DNA, and describe the structure of DNA • Assess the work of Watson, Crick, Wilkins and Franklin on DNA structure	Worksheet 3.1.9a; Worksheet 3.1.9b; Practical sheet 3.1.9	Quick starter; Interactive activity: Complete the sentences about DNA	Include an introduction to the passing on of genetic information, from 3.1.8

3.1.10	Learning about DNA	A simple model of chromosomes, genes and DNA in heredity, including the part played by Watson, Crick, Wilkins and Franklin in the development of the DNA model	• Identify that all plant and animal cells contain DNA • Explain why it is important for scientists to be able to extract DNA from cells • Analyse and evaluate the use of extracted DNA	Worksheet 3.1.10a; Worksheet 3.1.10b; Practical sheet 3.1.10; Technician's notes 3.1.10	Quick starter; Slideshow: Working with DNA – a look at DNA extraction and its use in forensics; Interactive activity: Drag the statements about DNA data to the correct group - ethical issue, scientific issue, or both	
3.1.11	Exploring human chromosomes	A simple model of chromosomes, genes and DNA in heredity, including the part played by Watson, Crick, Wilkins and Franklin in the development of the DNA model	• Identify that, at fertilisation, one chromosome in each pair comes from each parent • Explain how fertilisation results in each new individual being genetically unique • Explain how some genetic disorders arise	Worksheet 3.1.11a; Worksheet 3.1.11b; Practical sheet 3.1.11	Quick starter; Interactive activity: Reorder the sentences to describe the process of sexual reproduction in humans; Interactive activity: Drag the symptoms to the correct genetic disorder - Down's syndrome or cystic fibrosis; Slideshow: Chromosomal disorders: Explanation and examples	
3.1.12	Understanding cloning	A simple model of chromosomes, genes and DNA in heredity, including the part played by Watson, Crick, Wilkins and Franklin in the development of the DNA model	• Define cloning and describe some natural cloning processes • Explain how organisms may be artificially cloned • Explore ethical issues around artificial cloning • Compare and contrast asexual and sexual reproduction	Worksheet 3.1.12; Practical sheet 3.1.12; Technician's notes 3.1.12	Quick starter; Interactive activity: Reorder the sentences to describe how to clone a spider plant; Interactive activity: Reorder the sentences to describe how Dolly the sheep was cloned; Video	

3.1.13	Explaining extinction	Changes in the environment may leave individuals within a species, and some entire species, less well adapted to compete successfully and reproduce, which in turn may lead to extinction The importance of maintaining biodiversity and the use of gene banks to preserve hereditary material	• Identify changes which can cause a species to become extinct • Explain the use of gene banks to preserve hereditary material before a species becomes extinct • Analyse and evaluate theories of what caused the extinction of the dinosaurs	Worksheet 3.1.13	Quick starter; Hangman: Key vocabulary game	

Chapter 2: Our Health and the Effects of Drugs

Lesson	Lesson title	Overarching objectives	Learning objectives	CD-ROM resources	Collins Connect resources	Notes for two-year scheme
3.2.2	Exploring types of drugs	The effects of 'recreational' drugs (including substance misuse) on behaviour, health and life processes	• State examples of the four main groups of drugs • Describe the effects of different types of drugs on the body • Explain the effects of each type of drug on the body	Worksheet 3.2.2	Quick starter; Slideshow: Exploring types of drugs – information about the four main types of drug; Interactive activity: Drag the statements into the correct group - stimulant, depressant, painkiller or hallucinogen; Video	
3.2.3	Understanding the impact of smoking	The effects of 'recreational' drugs (including substance misuse) on behaviour, health and life processes The impact of exercise, asthma and smoking on the human gas exchange system	• Describe the effects of smoking on the body • Explain the risks of smoking on the body • Examine the link between smoking and cancer	Worksheet 3.2.3	Quick starter; Interactive activity: Match the component of tobacco to its harmful effect on the body	Include the four main groups of drugs from 3.2.2

3.2.4	Considering the dangers of cannabis	The effects of 'recreational' drugs (including substance misuse) on behaviour, health and life processes The impact of exercise, asthma and smoking on the human gas exchange system	• Describe the medicinal uses for cannabis • Describe the negative effects of cannabis on the body • Give a balanced argument about whether cannabis should be legalised	Worksheet 3.2.4	Quick starter; Interactive activity: Drag the statements into the correct group - for or against the legalisation of cannabis	
3.2.5	Understanding the effects of alcohol	The effects of 'recreational' drugs (including substance misuse) on behaviour, health and life processes	• Describe the short-term effects of alcohol on the body • Explain the long-term effects of alcohol • Suggest how alcoholism affects society	Worksheet 3.2.5	Quick starter; Interactive activity: Reorder the sentences to describe what happens to the body if a person continues to drink alcohol; Slideshow: Understanding the effects of alcohol: Information about alcohol and the dangers of long-term use; Video	
3.2.6	Exploring the effects of other drugs	The effects of 'recreational' drugs (including substance misuse) on behaviour, health and life processes	• Describe the effects of different drugs on the body • Compare the dangers of different drugs	Worksheet 3.2.6	Quickstarter; Interactive activity: Drag the drugs into the correct group - legal or illegal	

3.2.7	**Learning about addiction**	The effects of 'recreational' drugs (including substance misuse) on behaviour, health and life processes	• Define addiction • Describe how drugs affect the brain • Explain the effects of withdrawal on the body	Worksheet 3.2.7	Quick starter; Interactive activity: Drag the withdrawal symptoms into the correct group - emotional or physical; Hangman: Key vocabulary game	
3.2.8	Applying key ideas		• Extract information about the effects of alcohol and other drugs on drivers • Consider reasons why it is difficult to gather evidence about the effects of drugs • Use ideas and information about driving under the influence to form opinions and make reasoned arguments about solutions and punishments	Worksheet 3.2.8		
3.2.9	Understanding how diseases are spread	This part of this chapter is intended to embed and develop ideas from earlier in the Key Stage 3 course, including cell structure and function, micro-organisms and body systems.	• Describe how diseases are spread • Consider ways of reducing the spread of specific diseases	Worksheet 3.2.9	Quick starter; Slideshow: Preventing the spread of disease – a look at how infectious diseases are spread and how this can be prevented; Interactive activity: Match the way a disease is spread to the mechanism by which it can be prevented; Video	

3.2.10	Exploring the body's defences	This part of this chapter is intended to embed and develop ideas from earlier in the Key Stage 3 course, including cell structure and function, micro-organisms and body systems.	• Describe how the body resists infection • Explain the role of white blood cells in fighting infection	Worksheet 3.2.10	Quick starter; Interactive activity: Match the body barrier to the way it prevents microbe entry	
3.2.11	Comparing microbes	This part of this chapter is intended to embed and develop ideas from earlier in the Key Stage 3 course, including cell structure and function, micro-organisms and body systems.	• Describe the characteristics of different types of microbe • Recall examples of diseases caused by bacteria, viruses and fungi • Evaluate a model of a microbe	Worksheet 3.2.11; Practical sheet 3.2.11; Technician's notes 3.2.11	Quick starter; Slideshow: Types of microbe – a look at the features of fungi, viruses and bacteria; Video; Interactive activity: Drag the statements into the correct group - bacteria, virus or fungi	
3.2.12	Investigating the growth of bacteria	This part of this chapter is intended to embed and develop ideas from earlier in the Key Stage 3 course, including cell structure and function, micro-organisms and body systems.	• Describe what bacteria need to survive • Investigate bacterial growth in different conditions • Analyse bacterial growth data	Worksheet 3.2.12; Practical sheet 3.2.12; Technician's notes 3.2.12	Quick starter; Interactive activity: Order the surfaces based on the amount of bacteria present, from lowest to highest	

3.2.13	Understanding how antibiotics work	These topics are aimed at embedding and developing ideas from topics earlier in the scheme, including cell structure and function, and body systems, in the context of health	• Investigate the effect of antibiotics on bacteria • Explain how bacteria can become immune to antibiotics • Evaluate the impact of superbugs on our health	Worksheet 3.2.13; Practical sheet 3.2.13; Technician's notes 3.2.13	Quick starter; Interactive activity: Match the bacteria-killing chemical to its purpose; Slideshow: Antibiotics: A look at how antibiotics work and how you can test the effectiveness of an antibiotic	
3.2.14	Learning about vaccination	These topics are aimed at embedding and developing ideas from topics earlier in the scheme, including cell structure and function, and body systems, in the context of health	• Describe how vaccines were discovered • Explain how vaccines prevent a viral infection • Evaluate the risks involved with vaccination	Worksheet 3.2.14	Quick starter; Slideshow: Jenner, Fleming and Lister – a look at the famous scientists involved in the treatment or prevention of infectious disease; Interactive activity: Reorder the sentences to describe how a vaccination works; Hangman: Key vocabulary game	

Chapter 3: Obtaining Useful Materials

Lesson	Lesson title	Overarching objectives	Learning objectives	CD-ROM resources	Collins Connect resources	Notes for two-year scheme
3.3.2	Obtaining metals from ores	Earth as a source of limited resources	• Recognise how abundant common ores are in the Earth • Explain how ores are extracted from the Earth	Worksheet 3.3.2	Quick starter; Interactive activity: Drag the examples of substances to the correct group - native metals, non-native metals or ores; Video; Interactive activity: Place, in order, the metals from most to least reactive	
3.3.3	Understanding reactivity	The order of metals and carbon in the reactivity series Representing chemical reactions using formulas and using equations Thermal decomposition	• Use evidence to identify the reactivity series of metals • Represent reactions using formulas and equations	Worksheet 3.3.3; Practical sheet 3.3.3; Technician's notes 3.3.3	Quick starter; Interactive activity: Place, in order, the metals from most to least reactive; Slideshow: Reactivity of metals: Some examples	

3.3.4	Making use of displacement reactions	The order of metals and carbon in the reactivity series Representing chemical reactions using formulas and using equations Displacement reactions Conservation of mass changes of state and chemical reactions	• Represent and explain displacement reactions using formulas and equations • Make inferences about reactivity from displacement reactions	Worksheet 3.3.4; Practical sheet 3.3.4; Technician's notes 3.3.4	Quick starter; Interactive activity: Match the reactants to the products; Interactive activity: Place, in order, the elements from most to least reactive	
3.3.5	Using carbon to extract iron	The use of carbon in obtaining metals from metal oxides Representing chemical reactions using formulas and using equations Conservation of mass changes of state and chemical reactions	• Represent displacement reactions with carbon and metal oxides using formulas and equations • Explain how mass is conserved in the extraction of metals	Worksheet 3.3.5; Technician's notes 3.3.5	Quick starter; Interactive activity: Drag the statements to the correct group; Slideshow: The blast furnace – what goes in, and what comes out	
3.3.6	Extracting copper, lead and zinc	The use of carbon in obtaining metals from metal oxides Representing chemical reactions using formulas and using equations	• Explain how copper, lead and zinc are extracted from their ores • Calculate the yield of the extraction process	Worksheet 3.3.6; Practical sheet 3.3.6; Technician's notes 3.3.6	Quick starter; Interactive activity: Drag the statements to the correct group - reduction, oxidation or decomposition	Include the use of carbon to extract iron, from 3.3.5, and the impacts of extraction, from 3.3.7

3.3.7	Looking at the impact of metal extraction	The use of carbon in obtaining metals from metal oxides The production of carbon dioxide by human activity and the impact on climate Earth as a source of limited resources and the efficacy of recycling	• Describe the environmental impacts of metal extraction • Describe how recycling of metals reduces damage to the environment	Worksheet 3.3.7	Quick starter; Video; Interactive activity: Match the pollution to its effect; Slideshow: Recycling metals – a look at how we recycle metals, including e-waste; Hangman: Key vocabulary game	
3.3.8	Applying key ideas		• Extract ideas relating to how metals are used to protect each other • Apply ideas about the reactivity series, writing word equations for displacement reactions • Use ideas and information about particles to write balanced symbol equations and write detailed explanations of metal extraction processes			

3.8.9	Understanding exothermic reactions	Internal energy stored in materials Exothermic chemical reactions (qualitative) Comparing the starting with the final conditions of a system and describing increases and decreases in the amounts of energy associated with chemical compositions	• Describe examples of exothermic reactions • Explain the energy changes taking place during an exothermic reaction	Worksheet 3.3.9; Practical sheet 3.3.9; Technician's notes 3.3.9	Quick starter; Interactive activity: Reorder the sentences to describe what happens when substances react	
3.3.10	Comparing endothermic and exothermic reactions	Exothermic and endothermic chemical reactions (qualitative) Comparing the starting with the final conditions of a system and describing increases and decreases in the amounts of energy associated with chemical compositions	• Describe examples of endothermic reactions • Compare the energy changes during exothermic and endothermic reactions	Worksheet 3.3.10; Practical sheet 3.3.10; Technician's notes 3.3.10	Quick starter; Interactive activity: Drag the statements to the correct group - exothermic or endothermic change	Begin by introducing exothermic reactions, from 3.3.9
3.3.11	Explaining the use of catalysts	What catalysts do	• Describe what a catalyst is • Explain how catalysts work	Worksheet 3.3.11; Practical sheet 3.3.11; Technician's notes 3.3.11	Quick starter; Slideshow: Catalysis of hydrogen peroxide decomposition – a method to compare catalytic effectiveness; Interactive activity: Match the catalyst to its use	

3.3.12	Exploring ceramics and their properties	Properties of ceramics (qualitative)	• Describe what is meant by the term ceramic • Describe the properties of ceramics	Worksheet 3.3.12; Practical sheet 3.3.12	Quick starter; Interactive activity: Drag the statements to the correct group - ceramic, plastic or metal; Video	
3.3.13	Matching properties of ceramics to their uses	Properties of ceramics (qualitative)	• Explain how the properties of ceramics determine their uses	Worksheet 3.3.13	Quick starter; Slideshow: Ceramic materials – what are they, and where are they used?; Interactive activity: Place the materials, in order, from highest to lowest melting point	Include a discussion of properties, from 3.3.12
3.3.14	Exploring natural polymers	Properties of polymers (qualitative)	• Explain what a polymer is • Describe examples of natural polymers	Worksheet 3.3.14; Technician's notes 3.3.14	Quick starter; Interactive activity: Match the monomer to the polymer; Slideshow: Some natural polymers – examples	
3.3.15	Using human-made polymers	Properties of polymers (qualitative)	• Describe how human-made polymers are made in simple terms • Describe uses for human-made polymers	Worksheet 3.3.15	Quick starter; Video; Interactive activity: Reorder the sentences to describe how to make polypropene film; Interactive activity: Match the polymer to its use	

3.3.16	Explaining composites	Properties of composites (qualitative)	• Explain what is meant by the term 'composite' • Describe some uses of natural composites	Worksheet 3.3.16; Practical sheet 3.3.16; Technician's notes 3.3.16	Quick starter; Slideshow: Composites – examples of use; Interactive activity: Drag the statements to the correct group - human-made composites, natural composites, binder or reinforcer	May be merged with 3.3.17
3.3.17	Using human-made composites	Properties of composites (qualitative)	• Explain how human-made composites were developed • Describe the properties and uses of human-made composites	Worksheet 3.3.17; Practical sheet 3.3.17; Technician's notes 3.3.17	Quick starter; Interactive activity: Place the materials, in order, from least to most dense; Interactive activity: Matching metals and composites to their tensile strength:mass ratio; Hangman: Key vocabulary game	May be merged with 3.3.16

Chapter 4: Using our Earth Sustainably

Lesson	Lesson title	Overarching objectives	Learning objectives	CD-ROM resources	Collins Connect resources	Notes for two-year scheme
3.4.2	Understanding our atmosphere	The composition of the atmosphere	• Describe the composition of our atmosphere • Describe how the atmosphere has changed over time • Explain why the atmosphere has changed	Worksheet 3.4.2; Practical sheet 3.4.2; Technician's notes 3.4.2	Quick starter; Interactive activity: Drag the statements to the correct group - element in clean air, compound in clean air, or pollutant in air; Interactive activity: Match the components of air to the percentages present	
3.4.3	Exploring the effects of human activity	The production of carbon dioxide by human activity and the impact on climate	• Describe examples of human activity that cause air pollution • Explain the effects of smog, acid rain and damage to the ozone layer	Worksheet 3.4.3a; Worksheet 3.4.3b; Practical sheet 3.4.3; Technician's notes 3.4.3	Quick starter; Slideshow: Effects of air pollution – carbon dioxide, UV, ozone and acid rain; Video	May be merged with 3.4.4
3.4.4	Understanding the global warming debate	The production of carbon dioxide by human activity and the impact on climate	• Describe the effects of global warming • Explain the consequences of global warming for living things • Evaluate the arguments for human activity impacting on global warming	Worksheet 3.4.4	Quick starter; Interactive activity: Reorder the sentences to describe how we receive energy from the Sun; Interactive activity: Match the sea surface temperature to the year	May be merged with 3.4.3

3.4.5	Understanding how carbon is recycled	The carbon cycle	• Describe the carbon cycle • Explain how human activity increases the amount of carbon in the atmosphere • Explain what is meant by a 'carbon footprint'	Worksheet 3.4.5; Practical sheet 3.4.5	Quick starter; Interactive activity: Drag the statements to the correct group - decomposers or green plants; Slideshow – carbon cycle: releasing carbon dioxide; Interactive activity: Match the changes in the carbon cycle to the correct term	
3.4.6	Exploring damage to the Earth's resources	Earth as a source of limited resources and the efficacy of recycling	• Describe resources that the Earth provides • Explain how human activity limits these resources • Justify decisions about making changes to the environment	Worksheet 3.4.6	Quick starter; Interactive activity: Drag the statements to the most appropriate group - scars the landscape, pollutes the air, or damages or destroys natural habitats; Video	
3.4.7	Considering the importance of recycling	Earth as a source of limited resources and the efficacy of recycling	• Describe examples of recycling • Explain the benefits and limitations of recycling schemes • Compare the efficiency of recycling methods	Worksheet 3.4.7; Practical sheet 3.4.7; Technician's notes 3.4.7	Quick starter; Interactive activity: Place the materials in order, from the fastest to decompose to the slowest; Slideshow: Recycling issues – a look at some issues of recycling, including its problems; Hangman: Key vocabulary game	

3.4.8	Applying key ideas		• Extract ideas about damage to the Earth's resources caused by landfill sites and the importance of recycling • Apply ideas about recycling in nature and by humans to a new situation • Use ideas and information about global warming	Worksheet 3.4.8		
3.4.9	Understanding the structure of the Earth	The composition of the Earth The structure of the Earth	• Describe the layers of the Earth • Describe the characteristics of the different layers • Explain how volcanoes change the Earth	Worksheet 3.4.9a; Worksheet 3.4.9b	Quick starter; Interactive activity: Drag the statements to the correct group - inner core, outer core, mantle or crust; Video; Interactive activity: Reorder the sentences to describe how volcanoes form and erupt	
3.4.10	Exploring igneous rocks	The rock cycle and the formation of igneous, sedimentary and metamorphic rocks	• Describe how igneous rocks are formed • Explain how the pH of the magma affects the formation of rocks • Investigate the effect of cooling rate on the formation of crystals	Worksheet 3.4.10; Practical sheet 3.4.10a; Practical sheet 3.4.10b; Technician's notes 3.4.10	Quick starter; Interactive activity: Drag the statements to the correct group - volcanoes producing acidic magma or alkaline magma; Interactive activity: Match the cooling rate to the crystal size	

3.4.11	Studying sedimentary rocks	The rock cycle and the formation of igneous, sedimentary and metamorphic rocks	• Describe how sedimentary rocks are formed • Explain how fossils give clues about the past • Explain the properties of sedimentary rocks	Worksheet 3.4.11; Practical sheet 3.4.11a; Practical sheet 3.4.11b; Technician's notes 3.4.11	Quick starter; Video; Slideshow: Clues from fossils – what do fossils tell us?; Interactive activity: Reorder the sentences to describe the freeze-thaw process	
3.4.12	Using metamorphic rocks	The rock cycle and the formation of igneous, sedimentary and metamorphic rocks	• Name some examples of metamorphic rocks • Describe how metamorphic rocks are formed • Explain why metamorphic rocks are suited to their uses	Worksheet 3.4.12; Practical sheet 3.4.12; Technician's notes 3.4.12	Quick starter; Interactive activity: Drag the statements to the correct group - metamorphic or rocks of other types; Interactive activity: Match the materials before and after metamorphic change	
3.4.13	Understanding the rock cycle	The rock cycle and the formation of igneous, sedimentary and metamorphic rocks	• Describe the rock cycle • Explain how rocks can change from one type to another	Worksheet 3.4.13a; Worksheet 3.4.13b; Practical sheet 3.4.13; Technician's notes 3.4.13	Quick starter; Interactive activity: Place the processes of the rock cycle and the types of rock in the correct sequence, starting with 'Weathering and erosion'; Slideshow: Layers and folds – a look at how rocks are weathered; Hangman: Key vocabulary game	

Chapter 5: Motion on Earth and in Space

Lesson	Lesson title	Overarching objectives	Learning objectives	CD-ROM resources	Collins Connect resources	Notes for two-year scheme
3.5.2	Describing journeys with distance–time graphs	The representation of a journey on a distance–time graph	• Gather relevant data to describe a journey • Use the conventions of a distance–time graph • Display the data on a distance–time graph	Worksheet 3.5.2; Practical sheet 3.5.2; Technician's notes 3.5.2	Quick starter; Slideshow: Time-lapse photography – some examples; Interactive activity: Match the sentences about distance–time graphs for a car journey	
3.5.3	Exploring journeys on distance–time graphs	The representation of a journey on a distance–time graph Speed and the quantitative relationship between average speed, distance and time (speed = distance ÷ time)	• Interpret distance–time graphs to learn about the journeys represented • Relate distance–time graphs to different situations and describe what they show	Worksheet 3.5.3; Practical sheet 3.5.3; Technician's notes 3.5.3	Quick starter; Video; Interactive activity: Drag the statements into the correct group - those which describe acceleration, and those which do not	
3.5.4	Understanding relative motion	Relative motion: trains and cars passing one another	• Describe the motion of objects in relation to each other • Explain the concept of relative motion • Apply the concept of relative motion to various situations	Worksheet 3.5.4; Practical sheet 3.5.4; Technician's notes 3.5.4	Quick starter; Interactive activity: Order the relative speeds of the cars, from the fastest to the slowest	

3.5.5	Analysing equilibrium	Opposing forces and equilibrium: weight held by stretched spring or supported on a compressed surface Using force arrows in diagrams, adding forces in one dimension, balanced and unbalanced forces	• Analyse situations to identify the various forces that are acting • Explore static situations in which objects are held in equilibrium and the nature of the forces involved	Worksheet 3.5.5; Practical sheet 3.5.5; Technician's notes 3.5.5	Quick starter; Video; Interactive activity: Drag the statements into the correct group - those which describe forces in equilibrium, and those which do not	
3.5.6	Exploring motion and equilibrium	Opposing forces and equilibrium: weight held by stretched spring or supported on a compressed surface Using force arrows in diagrams, adding forces in one dimension, balanced and unbalanced forces Forces being needed to cause objects to stop or start moving, or to change their speed or direction of motion (qualitative only)	• Explore dynamic situations which may involve equilibrium • Apply ideas about equilibrium to a situation in which an object is moving	Worksheet 3.5.6; Practical sheet 3.5.6; Technician's notes 3.5.6	Quick starter; Slideshow: Forces in sky-diving – from free-fall to an open parachute; Interactive activity: Match the sentences about motion and equilibrium; Hangman: Key vocabulary game	

3.5.7	Applying key ideas		• Analyse distance–time data • Interpret data and apply ideas about motion to the context of ferry and train travel • Use ideas about forces, including equilibrium, to compare and explain various aspects of motion			
3.5.8	Understanding gravitational fields	Gravity force, weight = mass × gravitational field strength (g), on Earth g = 10 N/kg, different on other planets and stars	• Describe gravity as a non-contact force • Explore the concept of gravitational field and weight • Relate this concept to life on Earth	Worksheet 3.5.8	Quick starter; Video; Interactive activity: Order the weights, from largest to smallest; Slideshow: Mass and weight – an explanation of the two	
3.5.9	Applying ideas about gravitational fields	Gravity force, weight = mass × gravitational field strength (g), on Earth g = 10 N/kg, different on other planets and stars	• Apply the concept of gravity causing weight to other situations • Explore implications of varying gravitational field strength	Worksheet 3.5.9	Quick starter; Interactive activity: Drag the statements into the correct group - those which describe a strong gravitational field, and those which describe a weaker field	
3.5.10	Looking at motion in the Solar System	Gravity forces between Earth and Moon, and between Earth and Sun (qualitative only)	• Relate ideas about gravitational fields to the Sun–Earth–Moon system • Use these ideas to explain position and motion of these bodies	Worksheet 3.5.10	Quick starter; Interactive activity: Order the planets of the Solar System, from the furthest from the Sun to the nearest; Slideshow: Orbital motion – a look at how we put satellites into space	

3.5.11	Describing stars and galaxies	Our Sun as a star, other stars in our galaxy, other galaxies	• Describe the characteristics of a star • Relate our Sun to other stars • Explain the concept of galaxies and the position of our galaxy compared to others	Worksheet 3.5.11	Quick starter; Interactive activity: Match the statements about stars, galaxies and the Universe	
3.5.12	Explaining the effects of the Earth's motion	The seasons and the Earth's tilt, day length at different times of year, in different hemispheres	• Describe variation in length of day, apparent position of the Sun and seasonal variations • Compare these with changes in the opposite hemisphere • Explain these changes with reference to the motion of the Earth	Worksheet 3.5.12; Practical sheet 3.5.12; Technician's notes 3.5.12	Quick starter; Video; Slideshow: The effects of rotation and tilt – the Sun shines by day, and the moon shines by night … don't they?; Interactive activity: Reorder the sentences to explain the effect of the Earth's tilted axis	
3.5.13	Measuring distances in the Universe	The light year as a unit of astronomical distance	• Recall that the light year is used to measure astronomical distances • Explain the limitation of units such as km in describing astronomical distances • Describe a technique for measuring the distance to distant objects	Worksheet 3.5.13; Practical sheet 3.5.13	Quick starter; Interactive activity: Match the descriptions to the correct astronomical distances; Hangman: Key vocabulary game	

Chapter 6: Waves and Energy Transfer

Lesson	Lesson title	Overarching objectives	Learning objectives	CD-ROM resources	Collins Connect resources	Notes for two-year scheme
3.6.2	Making waves	Waves on water as undulations which travel through water with transverse motion; these waves can be reflected, and add or cancel – superposition	• Describe the movement of waves in water • Understand reflection of waves • Understand superposition of waves	Worksheet 3.6.2; Practical sheet 3.6.2; Technician's notes 3.6.2	Quick starter; Slideshow: Making waves – a look at the up and down movement of waves; Video; Interactive activity: Define the key terms about waves and wave motion	
3.6.3	Exploring light waves	The similarities and differences between light and waves in matter Light waves travelling through a vacuum; speed of light	• Describe light as travelling in waves • Understand the similarities and differences between water waves and light waves • Explain the frequency of a wave	Worksheet 3.6.3	Quick starter; Interactive activity: Drag the statements about light and sound into the correct group - true or false	
3.6.4	Explaining properties of light waves	The transmission of light through materials: absorption, diffuse scattering and specular reflection at a surface	• Describe how light passes through different materials • Understand how light can be absorbed by materials • Explain the difference between diffuse scattering and specular reflection	Worksheet 3.6.4; Practical sheet 3.6.4; Technician's notes 3.6.4	Quick starter; Slideshow: Windows, shadows and mirrors; Interactive activity: Order the materials, from the most transparent to the most opaque	

3.6.5	Using the ray model	Use of ray model to explain imaging in mirrors, the pinhole camera, the refraction of light and convex lens in focusing (qualitative); the human eye	• Describe the ray model of light • Explain how the direction of light rays can be changed • Explain how a pinhole camera and the eye work	Worksheet 3.6.5; Practical sheet 3.6.5	Quick starter; Interactive activity: Reorder the sentences to describe how light enters the eye	
3.6.6	Understanding energy transfer by light	Light transferring energy from source to absorber, leading to chemical and electrical effects; photo-sensitive material in the retina and in cameras	• Describe light as a way of transferring energy • Give examples of chemical and electrical effects when materials absorb light • Explain changes that happen when materials absorb light	Worksheet 3.6.6; Practical sheet 3.6.6; Technician's notes 3.6.6	Quick starter; Interactive activity: Drag the objects into the correct group - source of light, or reflector of light; Slideshow: Making use of light energy; Video	
3.6.7	Exploring coloured light	Colour and the different frequencies of light, white light and prisms (qualitative only); differential colour effects in absorption and diffuse reflection	• Describe how a spectrum can be produced from white light • Compare the properties of light of different frequencies • Explain how light of different wavelengths can be split and recombined	Worksheet 3.6.7; Practical sheet 3.6.7; Technician's notes 3.6.7	Quick starter; Slideshow: Separating and combining colours; Interactive activity: Order the colours of the spectrum, from the shortest wavelength to the longest; Slideshow: Explaining refraction; Hangman: Key vocabulary game	

3.6.8	Applying key ideas		• Extract ideas from the text about light and its properties • Use information about reflection and refraction to explain the formation of rainbows • Apply the ideas to explain the way CDs and DVDs can produce coloured light			
3.6.9	Understanding energy transfer and change	Energy as a quantity that can be quantified and calculated; the total energy has the same value before and after a change Comparing the starting with the final conditions of a system and describing increases and decreases in the amounts of energy associated with movements, temperatures, changes in positions in a field, in elastic distortions and in chemical compositions Using physical processes and mechanisms, rather than energy, to explain the intermediate steps that bring about such changes	• Describe the ways in which energy is stored • Describe the ways that energy can be transferred from one store to another • Explain that any change – physical or chemical – results in a transfer of energy	Worksheet 3.6.9	Quick starter; Interactive activity: Match the description to the type of energy store; Video	

3.6.10	Explaining thermal conduction and radiation	Heating and thermal equilibrium: temperature difference between two objects leading to energy transfer from the hotter to the cooler one, through contact (conduction) or radiation; such transfers tending to reduce the temperature difference: use of insulators	• Describe the warming and cooling of objects • Explain the relationship between energy transfer and temperature change • Compare the transfer of energy by thermal conduction and by radiation	Worksheet 3.6.10; Practical sheet 3.6.10; Technician's notes 3.6.10	Quick starter; Slideshow: Energy transfers around the home; Interactive activity: Drag the statements and descriptions into the correct group - energy transfer by conduction, or energy transfer by radiation	
3.6.11	Understanding energy transfers by fuels and food	Energy as a quantity that can be quantified and calculated; the total energy has the same value before and after a change Comparing energy values of different foods (from labels) (kJ) Other processes that involve energy transfer: metabolism of food, burning fuels	• Describe the use of fuels in the home • Explain that foods are energy stores and that the amount stored can be measured • Explain that energy is not a material and can be neither created nor destroyed	Worksheet 3.6.11	Quick starter; Interactive activity: Drag the units into the correct group - unit of energy, or not a unit of energy; Video	
3.6.12	Comparing rates of energy transfer	Comparing power ratings of appliances in watts (W, kW) Comparing amounts of energy transferred (J, kJ, kW hour)	• Describe what is meant by 'rate of energy transfer' • Recall and use the correct units for rate of energy transfer • Calculate quantities of energy transferred when change happens	Worksheet 3.6.12	Quick starter; Interactive activity: Match the calculated quantity to the correct description	

3.6.13	Looking at the cost of energy use in the home	Comparing power ratings of appliances in watts (W, kW) Comparing amounts of energy transferred (J, kJ, kW hour) Domestic fuel bills, fuel use and costs	• Describe the information a typical fuel bill provides • Explain and use the units used on a fuel bill • Explain how the cost of energy used can be calculated	Worksheet 3.6.13	Quick starter; Interactive activity: Match the descriptions to the numbers and units found on household electricity bills; Hangman: Key vocabulary game	

Teaching route: Cells

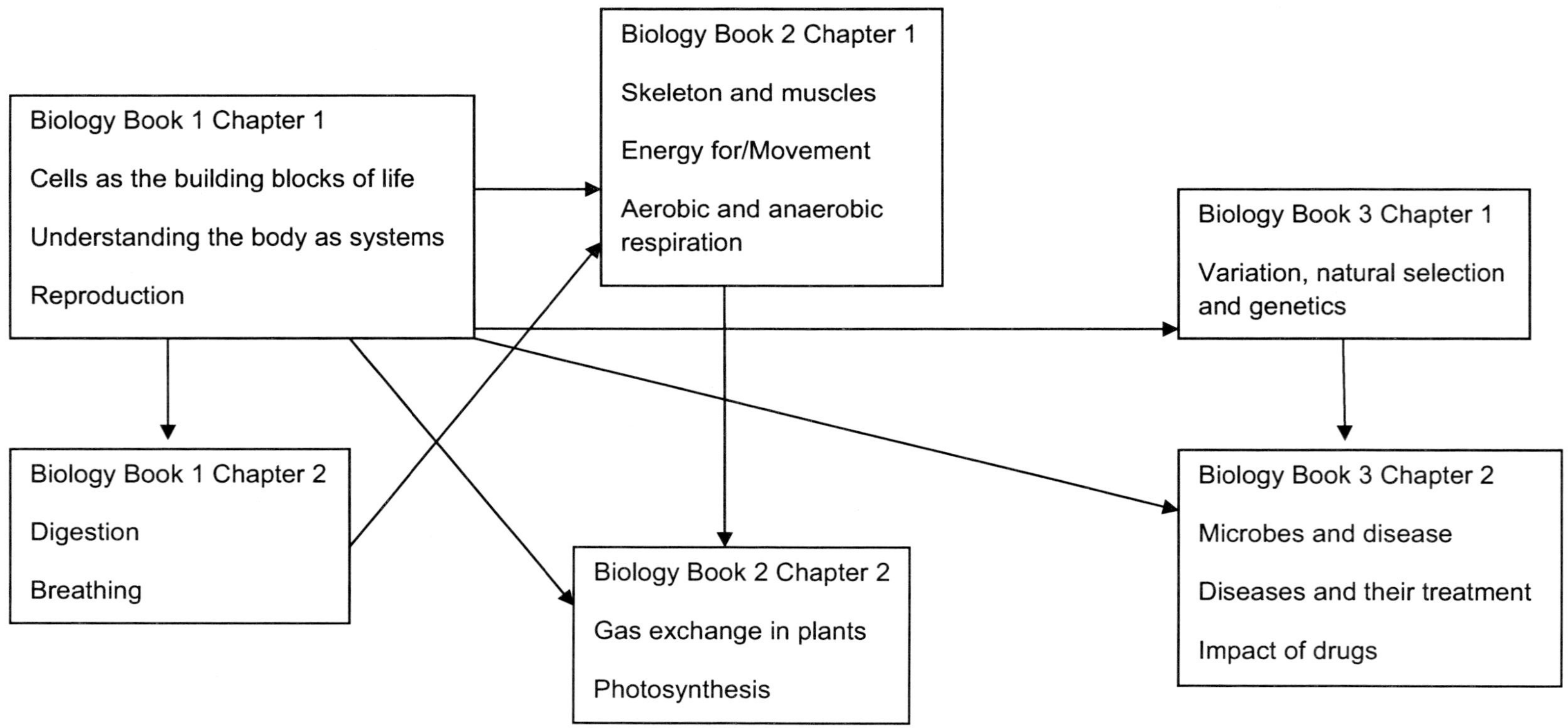

Teaching route: Energy

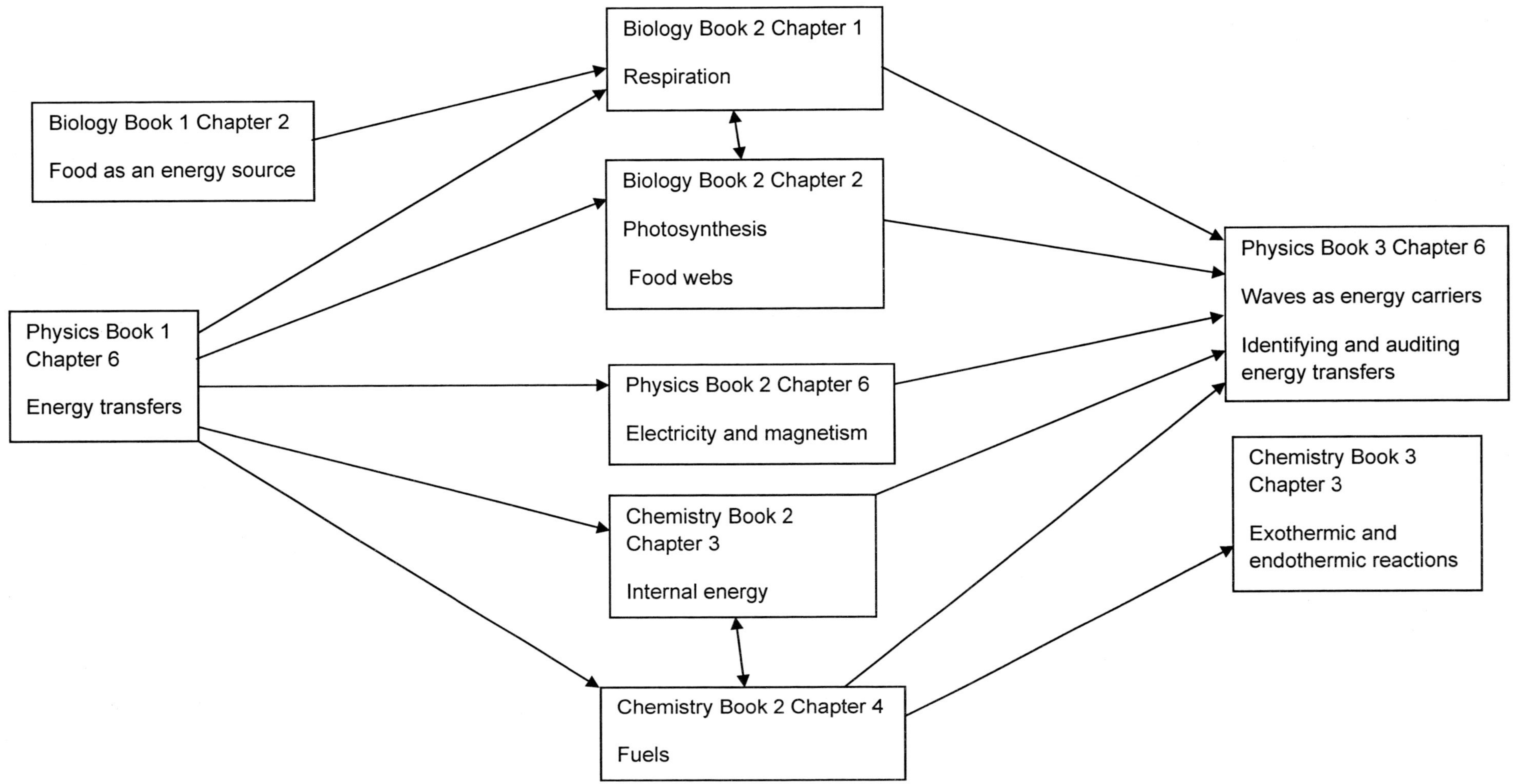

Teaching route: Forces

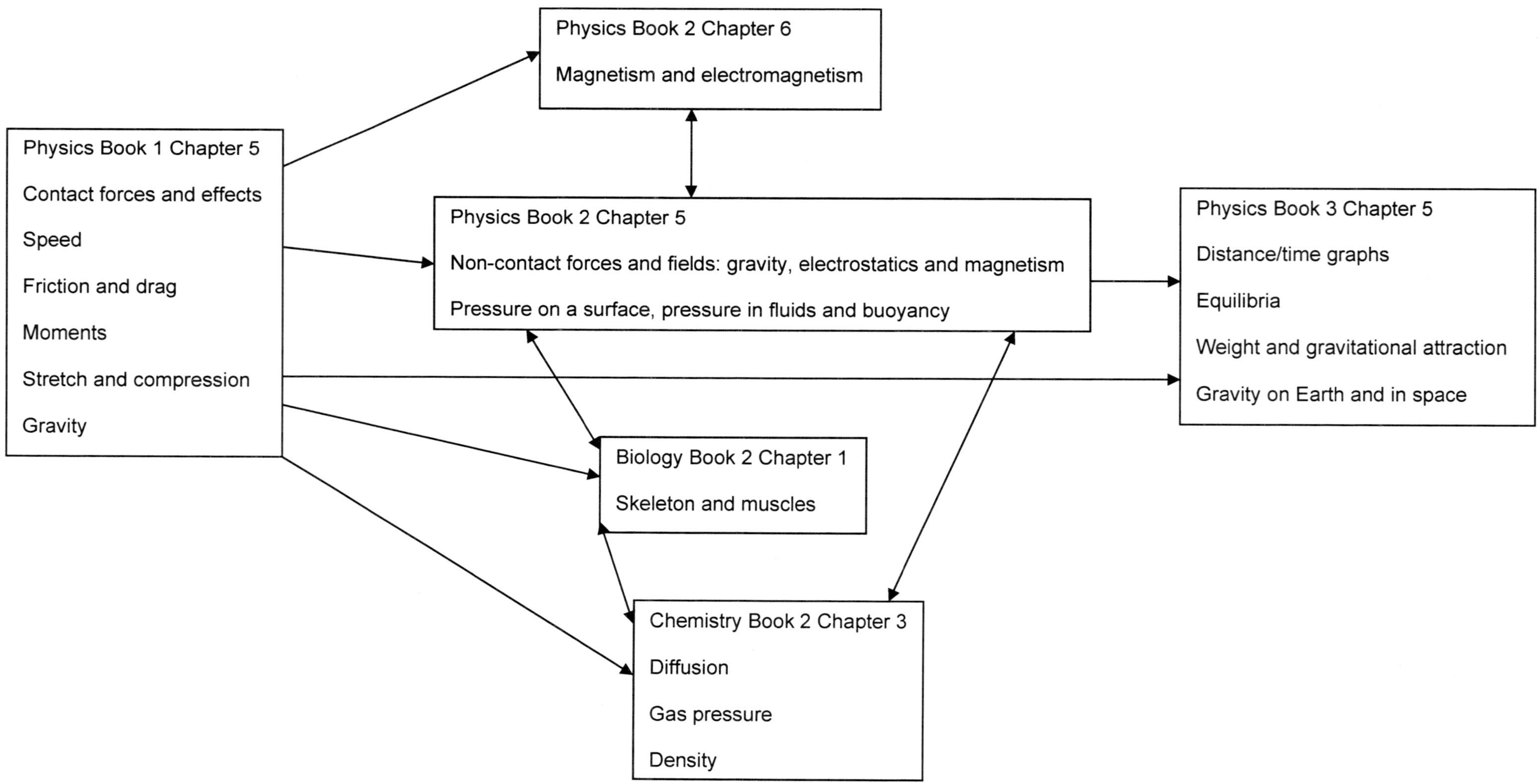

Teaching route: Particles and elements

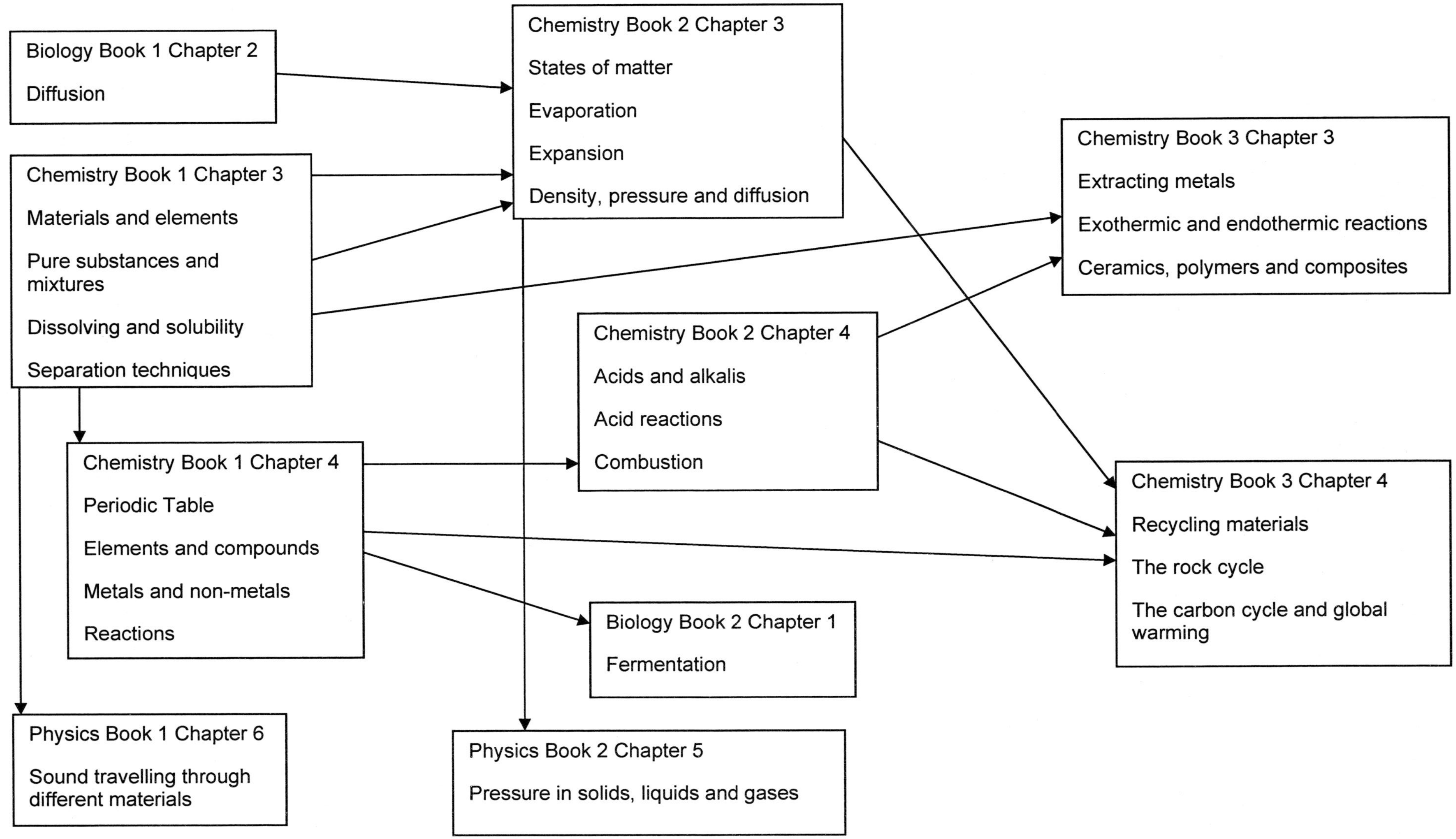

Teaching route: Interdependence

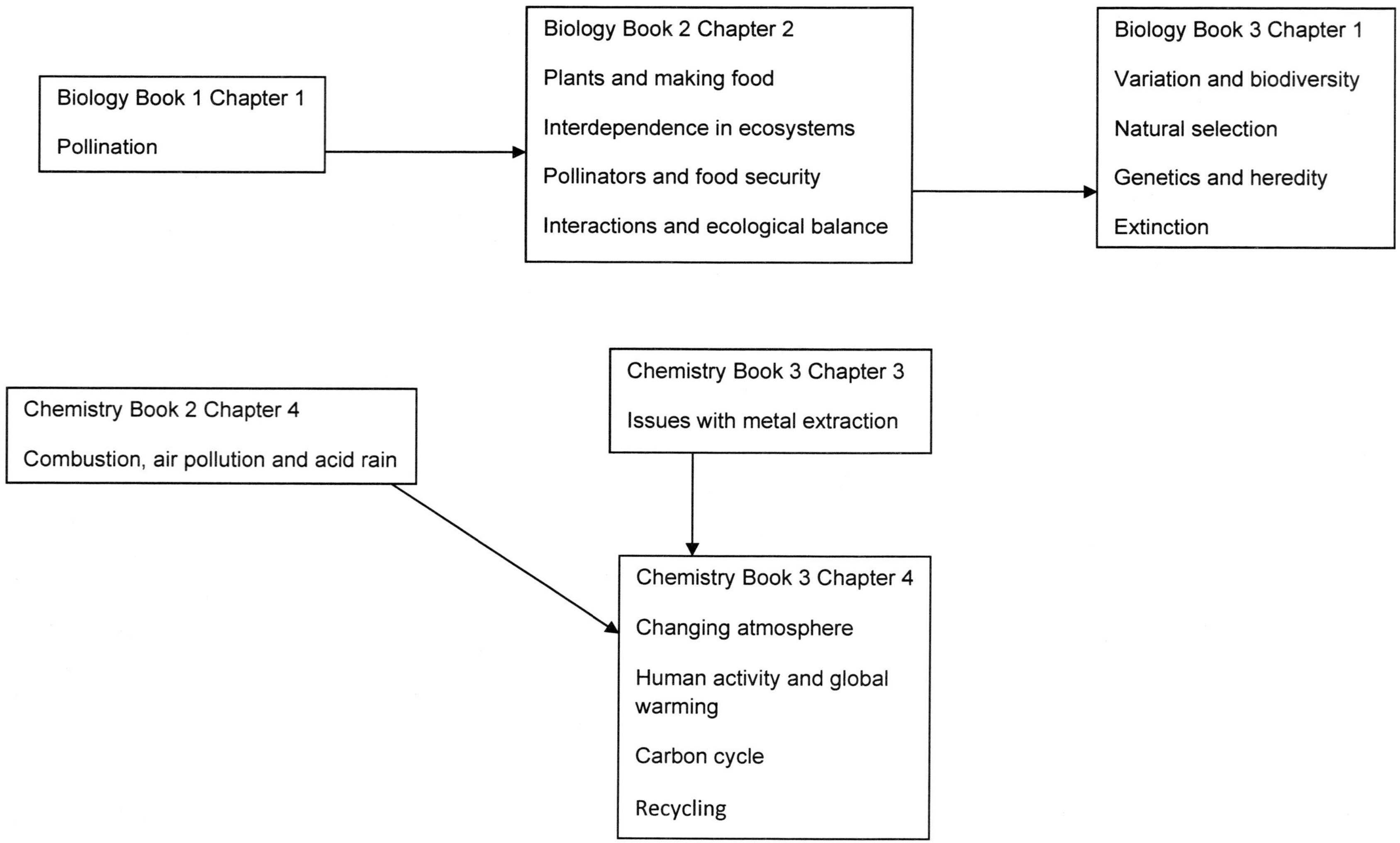

Notes

Notes

Notes

